THE YEAR
OF THE CRAB

Illustrations by Robert Jon Golder

W · W · NORTON & COMPANY · NEW YORK · LONDON

THE YEAR
OF THE CRAB

MARINE ANIMALS
IN MODERN MEDICINE

WILLIAM SARGENT

The text of this book is composed in ITC Garamond Light, with dis-
play type set in Romana. Composition and manufacturing by
The Maple-Vail Book Manufacturing Group.
Book design by Debra Morton Hoyt.

First published as a Norton paperback 1988

Library of Congress Cataloging-in-Publication Data

Sargent, William, 1946–
The year of the crab.

1. Marine fauna.   2. Limulus polyphemus.   3. Zoology,
Experimental.   4. Medicine, Experimental.   I. Title.
QL121.S27   1987   591.92   86–12474

ISBN 0-393-30540-6

W. W. Norton & Company, Inc.
500 Fifth Avenue, New York, N. Y. 10110
W. W. Norton & Company Ltd.
37 Great Russell Street, London WC1B 3NU

3 4 5 6 7 8 9 0

To my wife Kristina,
who more than anyone believed in this book.

CONTENTS

CONTENTS

PART THREE: AUTUMN
Marine Animals and the Human Brain

PART FOUR: WINTER
The Future

To Woods Hole

Nestled on the southern edge of that geological curiosity we call Cape Cod rests a unique community: part New England fishing village, part summer resort, and part world-famous marine science center. I have spent much of my life orbiting back to Woods Hole to rekindle an excitement first sparked aboard one of its research vessels.

I was a callow Harvard undergraduate who had been offered a job as a research assistant on a cruise to Africa, South America and the Baltic.

The first night I stood on the bow of the Atlantis II, watching as snowflakes swirled out of an inky sky, danced silver in the beam of our overhead light, and fell silently into the cold waters of the Atlantic. A few days later the deep green of the Atlantic gave way to the turquoise blue of the Gulf Stream. I watched dolphins trail wakes of phosphorescence as they rode the bow wave of the ship that was to be my home for the next six months.

During that time I spent long hours setting plankton nets, retrieving water bottles, and peering through microscopes at planktonic creatures of the most exotic colors and bizarre forms. Some nights I would haul a mattress down to the recesses of the ship, where the bow observation chamber allowed me to watch diaphanous orbs of plankton loom out of the darkness like silent spaceships against a star-flecked void.

But we were not entirely cut off from land. One day we trekked across the rugged terrain of Saint Peter and Saint Paul's Rocks, a jagged outcrop of the mid-Atlantic ridge that rises only fifteen feet above the ocean midway between South America and Africa. Darwin traversed these same rocks and observed the ancestors of the same sooty and noddy terns. On another occasion we stayed up all night carousing through the streets of Dakar with a group of drunken Soviet oceanographers on their maiden cruise aboard the *Akademik Kurchatov.*

But it was the day-to-day routine—working on the decks, playing endless games of Scrabble, getting to know people—that was to irrevocably change my life. The experience awakened my resolve to return to Woods Hole to continue the adventure.

Today I live in Woods Hole overlooking the waters that surge through the "hole" between Naushon Island and the mainland. At every change of tide the currents reverse, almost pulling the largest buoys underwater. It makes this one of the most treacherous passages on the East Coast, but a boater's nemesis is often a naturalist's delight.

Naturalists started coming to Woods Hole during the

late 1800s. At that time Woods Hole was a wealthy summer resort, rivaling Newport and Bar Harbor in opulence. Bar Neck Road, nicknamed "Banker's Row," included the summer homes of many of the country's most wealthy families.

In 1871 Spencer Baird visited the wealthy watering hole. An astute scientist with considerable political finesse, Baird parlayed a one-hundred dollar research grant into a new federal agency, the United States Fisheries Commission, with its first permanent laboratory in Woods Hole.

Had it not been for Spencer Baird's success, Woods Hole might still be a stuffy New England resort. Tallulah Bankhead had *that* Woods Hole in mind when she informed several writers that they could "Take your Nantucket and shove it up Woods Hole." Of course she might have had a jaundiced view, as she was being escorted up the gangplank of the ferry by the Nantucket board of selectmen, who disapproved of her late-night parties.

Fortunately for naturalists, the fisheries laboratory proved to be a magnet for other institutions. The Marine Biological Laboratory was added in 1888 and the Woods Hole Oceanographic Institution was established in 1930. Today the findings of Woods Hole's scientists are known throughout the world.

Despite Woods Hole's international reputation, it maintains a small-town flavor. Neckties are suspect and parking spaces are few. The postmaster is the first to know, and to tell, who has been elected to the National Academy of Sciences. In the evening Nobel prizewinners still amble down to the docks, buckets in hand, to retrieve some squid for dinner, while research assistants tell tales of the bluefish that got away.

Woods Hole scientists spend several months a year

exploring the deepest reaches of the oceans, collecting water samples and teasing apart the tiniest organelles of such seemingly humble creatures as horseshoe crabs, lobsters, and squid.

For the past few years I have pursued a personal quest to understand horseshoe crabs. The quest has led me on a fascinating odyssey of discovery, an odyssey which in many ways has recapitulated the history of modern biology. Horseshoe crabs were first described by the intelligent French explorer Samuel de Champlain in 1605, when natural history was in vogue. The horseshoe crab remained a zoological curiosity until Darwin transformed modern thought with what we now call the theory of evolution. During the turn of the century, several seaside laboratories were established to elaborate on the finer points of the theory. Embryology seemed to be the key, and the horseshoe crab's ancient trilobite larvae were of prime interest.

Following the Cartesian method of reductionism, biologists began to investigate ever smaller details in order to unravel the greater mysteries of life. They moved from the whole organism to the embryo, to the cell, to DNA and its constituent molecules.

During the process horseshoe crabs played a significant role. The long single cell of the horseshoe crab's optic nerve became a model for the function of nerves and helped establish the field of neurobiology. Most of what we know about vision was elucidated from Dr. H. Keffer Hartline's Nobel prizewinning work on lateral inhibition by cells in the horseshoe crab's compound eye.

The primitive large amoebocyte cell of the horseshoe crab played a crucial role in unraveling the evolution of antibodies in higher mammals. During the past decade, biologists have relied on this former basic research to make dramatic breakthroughs in applied biomedicine. Gene-splicing companies presently hold center stage, but right beside them and just a little bit ahead are, again, horseshoe crabs. A major biomedical industry based on the manufacture of processed horseshoe crab blood has already saved hundreds of lives from bacterial diseases and may, in turn, threaten the future of the very animals on which it depends.

My curious quest to understand horseshoe crabs has given me the unique privilege of looking over the shoulders of numerous Woods Hole scientists. I have watched as they input their data, process their information, and arrive at new theories about the natural world. Within days of the announcement of a new discovery at Woods Hole, its implications will be discussed at the Soviet Ministry of Fish Industry, fed into Chinese computers, debated at the *Stazione Zoologica di Napoli,* and incorporated into theories about the evolution of the planets at the Jet Propulsion Laboratory in Pasadena, California.

Like the developing dendrites of an infant's brain, this global network of research centers gathers information, processes data, and develops new ideas to alter our concepts of the world, the oceans, and ourselves. Woods Hole remains a small but vital part, a synapse or ganglion perhaps, in the ongoing and mysterious evolution of ideas that so preoccupies our species and gives such meaning to our existence. For this reason I dedicate this book to a village that has given me so much. I hope that I will be successful in passing some of it on to others.

ACKNOWLEDGMENTS

Books have a way of writing themselves. Like a molting horseshoe crab, this book has gone through many metamorphoses. The original book was a chronicle of my personal experiences studying horseshoe crabs. After a second editor and a new publishing company, it became a book about the natural history of horseshoe crabs; after a third editor it became a book about how marine animals are used in modern biomedical research.

As the book matured it drew me deeper and deeper into a rich store of research literature. I spent long hours wading through journals and medical textbooks. Along the way I was aided by many thoughtful people who explained their work, read my manuscript, kept me on the right path, and coaxed me forward. I would like to thank some of them here.

First, I would like to thank the crabs themselves, for they drew me along. My original intent was simply to raise horseshoe crabs. Other writers have had similar romantic notions—write in the morning and pursue some simple-minded pastoral occupation in the afternoon. Thoreau had his peas, E. B. White had his pullets; why couldn't Sargent have his horseshoe crabs?

ACKNOWLEDGMENTS

Of course the endeavor proved to be neither simple nor romantic. I spent long hours up to my keister in cold water and crabs trying to keep them alive. I had run-ins with shellfish wardens, town fathers, and narcotics agents. Can you imagine trying to explain to a federal agent that you are in your wetsuit, skulking about the marshes before sunrise in order to keep your horseshoe crabs from getting wet? Even the admiral of the Office of Naval Research did not share our enthusiasm for arachnids when one of them happened to get stuck in the main drain of his research facility.

But several people were very helpful and supportive. Tom Gedaminski, a student of mine from Boston College, put in long hours at minimal pay. My wife Kristina Lindborg could always be counted on to mend nets and recapture runaway crabs. Phil Coates, director of the Massachusetts Division of Marine Fisheries, and John Farrington from the Woods Hole Oceanographic Institution, were both interested and supportive of our work.

In Woods Hole my wife suggested that there was a book to be written about horseshoe crabs. Many people at the Marine Biological Laboratory in Woods Hole were supportive and helpful. Dr. Jella Atema allowed me to sit in on his animal behavior course. He, Bruce Johnson, and Paola Borroni helped answer many of my questions.

The staff of the MBL library was invaluable. The library is one of the wonders of the civilized world. It is open twenty-four hours a day, seven days a week, except Christmas. It operates entirely on the honor system. Anyone can sign out a book for as long as they need it, as long as the library can reach them in case someone else wants the book. A few years ago, in an effort to run things in a more efficient fashion, the library put out a general call for people to return their MBL books. The response of one scientist was typical. "Why that's absurd. The library isn't large enough to house all the MBL books we have at home."

Carl Schuster, Louis Liebowitz, Ted MacNichol, Alan Fein, Ed Kravitz, Betsy Bang, Peter Hartline, and Robert Barlow helped with their reminiscences and read some chapters. With great trepidation I presented my first draft to Paul Gross, director of the MBL. His positive comments spurred me on. Illustrator Bob Golder put up with the general mania as the book drew to a close, and his excellent illustrations bring life to the text.

Jim Mairs, a vice-president and editor at W. W. Norton and Co., Inc., somehow convinced his fellow editors and sales staff that they could sell a book about horseshoe crabs. My copy editor Patricia Peltekos and I had numerous spirited, yet helpful, conversations on the margins of the final draft.

My uncle, Dr. Sargent Cheever, aided in many ways. Had I known him earlier in life I might very well have pursued medicine rather than writing. E. O. Wilson and Lewis Thomas have been embarassingly kind in their praise of my books. Special thanks to my parents and family, who have been unswerving in their support of my rather unorthodox career.

PART ONE

SPRING

The Animals

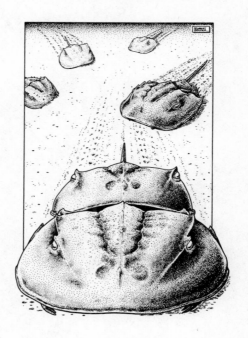

1

Spring Arrives on the Continental Shelf

The dark hulk of a lobster pot drifts down through the stygian gloom. It plunges into the sediments and a small cloud of ooze mushrooms off the ocean floor. The continental shelf returns to the cold quiet of early spring.

Yet there is life beneath these sediments. A large female horseshoe crab, *Limulus polyphemus,* is preparing to reenact a ritual that has persisted for eons. Entombed within the gentle grasp of the partially consolidated sediments, she has overwintered for the last five months. Now, without the benefit of changing light, temperature, or any other environmental cue, she is about to end her vigil.

Deep within the recesses of her primitive nervous system, a biological clock has kept time throughout the long dark winter. Faint electrical discharges have silently marked the seconds, minutes, days, and months until the time of her vernal migration. Her brain now signals that the allotted time has elapsed: her departure is at hand.

Gradually she hefts her heavy carapace from beneath the substrate and starts her fifty-mile migration across the North American continental shelf, the sandy flank formed by the accumulation of sediments eroded from the continent's interior. In two months' time she will reach the mouth of the bay where she will lay her eggs and spend the summer.

Farther to the east, a long line of lobsters, *Homarus americanus,* crawl out of the head of one of the canyons that slice through the underwater escarpment of the continental shelf. The lobsters have wintered in burrows that honeycomb the canyon walls like the nests of swallows along a riverbank. Their burrows form "pueblo" communities housing eelpout and haddock, redfish and hake. Billions of delicate anemones festoon the canyon walls and miasmas of euphausid shrimp swarm through their swaying tentacles.

Like the horseshoe crab, the lobsters are starting their own migration, which will take them from the deep canyon to their inshore summering grounds, 150 miles across the continental shelf. They will feed on acre upon acre of spiny-skinned sea urchins, and crawl through the burrows of tilefish, some thirty feet in diameter and six feet deep.

Above the horseshoe crabs and lobsters, another ancient migration is taking place. Vast schools of fleshy squid, *Loligo pealei,* are slowly swimming through the murky waters. These mollusks have spent the winter in the deep water off the continental shelf. As the water temperature rises, they form dense schools, which are now migrating toward the coast. There the squid will partake in one of the strangest mating rituals of the animal kingdom.

For now, they migrate slowly. Bands of color undulate down the lengths of their bodies and their large, intelligent eyes scan the water in search of prey. Occasionally a large fish darts through the school and a hundred fusiform bodies shoot outward, forming a starburst pattern against the dark green waters of the Atlantic Ocean.

The squid migrate through swirling clouds of phytoplankton (tiny floating plants), whose presence signifies that the spring overturn has occurred, a phenomenon that makes spring underwater as dramatic as that on land. For several weeks, the lengthening days of sunlight have warmed the surface waters of the ocean, which have become more dense than those that lie below. Eventually the surface waters plunge through the underlying waters to the ocean floor. This overturn stirs up massive clouds of nutrients that had accumulated throughout the winter from the decay of plants and animals.

The upwelling of nutrients has fertilized the oceans, triggering blooms of phytoplankton, upon which the entire ocean food chain depends. They are followed by a blooming profusion of zooplankton, microscopic floating animals, which are now busily grazing on the phytoplankton.

Soon the waters will be replete with fish eggs, larval fish, and adults. It is all part of a well-orchestrated unfolding of life, finely attuned to the rising water temperature, increasing daylight, and renewal of nutrients. Each species responds to a particular cue at just the correct time when food is most abundant and fecundity best rewarded. By late spring, every quart of this water will swarm with millions of planktonic organisms.

On the surface, cold gusts of arctic air still roil the waters and white caps dance and glitter in the northern sun. A pair of stormy petrels flit over the waves, occasionally, stalling to pluck a choice morsel of plankton from the crest of a sun-flecked wave.

Not far away a fisherman watches the petrels from the stern of a lobster boat. He has just put in the first lobster set of the season. He takes off his gloves, wipes his brow and gazes toward the east, where a faint rose hue advances across the morning sky. The wind carries the sweet smell of spring. It is the tangy odor of phytoplankton that turns the ocean green with their abundance.

The seasonal cycle has begun again. The earth has swept along its preordained path so that the Northern Hemisphere now tilts toward the the sun's warming rays.

How many times have the horseshoe crabs' ancestors witnessed this annual renewal of life? Twelve thousand years ago this area was above water. Paleo-Indians came here to collect horseshoe-crab tails for their arrowheads. They used these meager weapons to hunt woolly mammoth beneath the face of the Wisconsin glacier, which was slowly retreating across North America. Two hundred million years ago, this area was part of Africa, a narrow margin of land that fractured when the continents split and started their ponderous migration outward.

During their 350 million years, horseshoe crabs have witnessed the construction and destruction of mountains, the flooding and receding of oceans, and the appearance and evolution of new species. Yet throughout the horseshoe crabs' impressively lengthy existence, the shallow waters surrounding the continents have remained—a changing yet durable environment.

Adapted to this environment from early in the history of life, horseshoe crabs have had little reason to evolve. Safe in the fitness of their particular form and behavior, these animals are built for survival.

2

The Paleozoic Crucible

> When you were a tadpole and I was a fish
>   In the Paleozoic time;
> And side by side on the ebbing tide
> We sprawled through the ooze and slime,
> Or skittered with many a caudal flip
> Through the depths of the Cambrian fen,
> My heart was rife with the joy of life,
> For I loved you even then.
>   —Langdon Smith, "Evolution"

It is a warm sunny day in the midst of the Paleozoic era, 300 million years ago. Beneath the shallow seas, flower-like crinoids sway in the gentle currents, large-eyed cephalopods propel themselves backward through the water, and myriads of trilobites scuttle across the ocean floor.

There are thousands of species of trilobites, quixotic creatures that have evolved curious structures and behaviors. Some swim on their backs, some burrow in the substrate, and others scoot across the sandy bottom. Many trilobite species have evolved life's first high-resolution eyes, while others remain blind.

By the end of the 350 million-year-span of the Paleozoic, however, trilobites will disappear and will be replaced by horseshoe crabs.

Curiously, the underwater world of the Paleozoic was not strikingly different from that of the present. Corals grew in profusion, worms slithered across the substrate, and jellyfish pulsed through sun-dappled waters.

But the terrestrial environment was quite different. While animals dominated the oceans, plants held sway on land. The continents were covered with a verdant landscape of giant tree ferns, club mosses, and occasional conifers. Animal species were few. Some dragonflies the size of dinner plates hovered over spiders, scorpions and four-inch cockroaches, but only these few ancient forms were present.

The Paleozoic was arguably the most significant and revolutionary period in the history of evolution. It was during this ancient era's brief span that most of the basic designs of life were first established.

Geologists divide the 4.6 billion years of the earth's history into two eons: the eon of hidden life, the Cryptozoic, and the eon of evident life, the Phanerozoic, when obvious fossils appear in the geologic record.

The Phanerozoic eon is further divided into several broad eras: the Paleozoic ("ancient life") known for its invertebrates; the Mesozoic ("middle life") known for its dinosaurs; and the Cenozoic ("recent life") known for humans and a few other mammals.

Evolution proceeds through periods of explosive revolution, relative quiescence, and dramatic collapse. True to form, the Paleozoic opened with the revolutionary Cambrian explosion, a flowering of life-forms unlike any that came before or after.

The Paleozoic was preceded by the Cryptozoic or Precambrian, a ponderous eon when blue-green algae appeared, then lay dormant for the better part of 2.5 billion years. The most significant event of this time was the "invention" of sex.

Formerly organisms reproduced by cloning exact copies of themselves, a tedious process that allowed for little change. Evolution could only proceed through the rare event of a chance mutation. The evolution of sex changed all that. Through sexual reproduction, organisms could exchange genetic material encoded with its collection of mutations. The result was that evolution proceeded more rapidly. This, the first and most significant "sexual revolution," was a necessary precursor to the dramatic changes to come.

There were undoubtedly other significant changes that led to the Cambrian explosion. For instance, the blue-green algae had built up a rich atmosphere of oxygen; the ocean temperature had cooled and its chemistry had altered so that animals could extract minerals to build shells. Some of these early organisms started to crop the dominant algae, making room for newer forms. This was a time when the planet itself evolved from a lifeless sphere to a synchronous body of animate and inanimate parts, similar to the structure of a living cell.

The explosion of life during the Cambrian marked the birth of the fertile Paleozoic era; during its short span, most of the major phyla of animals appeared and evolved the designs, forms, and behaviors that still dominate life on earth.

Suddenly, during the Permian, the Paleozoic came to a dramatic close. The land masses scattered across the planet coalesced into the supercontinent of Pangaea. It was a catastrophe for the newly evolved life forms. The shallow water environment surrounding each mini-continent was reduced by fifty percent.

Ocean ridges subsided. The ocean level dropped, leaving large areas of the continental shelf exposed. The shallow water environment of the shelf almost disappeared and with it went the multitudes of animals which had flourished during the Paleozoic. By the end of the great dying, over fifty percent of all marine invertebrates had disappeared.

Three major invertebrate designs—the flowerlike crinoids, the large-eyed cephalopods, and the scuttling trilobites—dominated the Paleozoic shallows. From these Paleozoic ancestors evolved our modern forms: from the crinoids emerged the sea urchins; the cephalopods evolved into the squid; and the ancestoral trilobite developed into the lobster and horseshoe crab. These three ancient forms survived because of their ability to adapt to a hostile environment. They solved the problems of getting food, avoiding predators, and reproducing their kind. It is for this reason that modern biomedical researchers turn to these four animals to unravel the secrets of nerves, muscles, blood, and vision.

Every time that you go to the hospital for a major procedure, your recovery very likely owes its success to basic research that has been carried out on one of these animals. It is their ancient heritage that has made them such crucial allies in our quest to understand life and reduce disease.

3

The Night of the Squid

On Capistrano they await the swallows. In Strasbourg they await the storks. In Woods Hole we wait for the return of the squid.

—Anonymous Biologist

It is early May. A broad plume of warming waters advances up the coast of North America, bringing with it a majestic underwater migration of returning species. Within the rips and runnels of the offshore bars, a living river of animals returns to the shores of the East Coast. Silver multitudes of shad and alewives riffle the ocean surface while rays glide and crabs crawl along the bottom. As the water temperature climbs above forty degrees Fahrenheit, lobsters and horseshoe crabs return from their offshore wintering grounds.

Squid have also completed their inshore migration from the waters of the continental slope. They have covered approximately one hundred miles in less than a month, and now millions of the fleshy mollusks congregate over patches of sandy bottom from Cape Cod to Cape Hatteras. These are the long-finned squid, *Loligo pealei,* whose scientific name honors two naturalists centuries apart. Pliny the Elder gave this squid its genus name, *Loligo,* and a later scientist honored the Philadelphia philanthropist and naturalist, Rubens Peale, son of the American painter, Charles Willson Peale.

By night, the feeble rays of the new moon filter through writhing multitudes of the foot-long creatures. The slightly larger males dart through the school trying to herd individual females away from the pack. Some females are more in demand and the males must dart, grapple, and occasionally bite their opponents in order to claim their prize.

A male separates a female from the school. Now they swim side by side, the male holding one of his long median arms aloft in an elegant "S"-shaped courting curve. Chromatophores flare along his head and body, and an intense reddish brown spot pulsates between his eyes. When another male approaches, this spot intensifies and dark splotches appear along the side of his body nearest to the challenging male.

Now the male glides below the female and attempts to grasp her around the middle. She pushes his arms away and darts aside. Again they swim side by side through the mass of frenzied squid. Again he glides beneath her and grasps her body. This time she does not resist. His arms grope forward toward her mantle opening. Holding her with nine strong tentacles, he reaches into his own mantle and withdraws a bundle of spermatophores from his muscular penis. With one quick motion he plunges the spermatophore bundle deep into her mantle where it

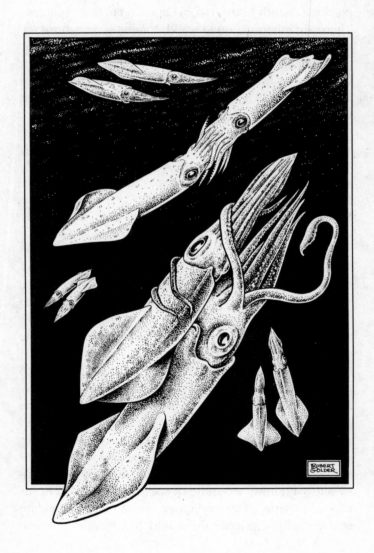

ejaculates and tiny reservoirs of sperm adhere to the mantle wall surrounding her oviduct. Within five to twenty seconds their mating is over.

Now the female turns her attention to spawning. She reaches into her mantle and withdraws an egg capsule encased within a thick mass of sperm-impregnated jelly. Holding the capsule before her, she swims directly toward the bottom, ignoring the advances of her still amorous partner. She locates a fucus-covered rock and plunges into the fronds of the leafy brown alga. With a series of twists and turns, she entwines the egg capsule into the fronds and flushes the eggs with her siphon before darting back to her suitor.

Contact with sea water hardens the egg jelly. The sight of the new mass elicits renewed frenzy among the other mating squid. One by one other females mate and swim down to deposit their egg capsules until a mass of pencil-length egg capsules gently sway in the tidal currents. Later in the season trawlers will dislodge many of the egg masses and they will wash up on the shore to confound beach walkers.

By dawn, the mating ritual has ceased. Many adults have succumbed to the thousands of predatory fish that cruised through the mating squid, gorging on their succulent bodies. Other adult squid have died from exhaustion. On the West Coast all the adults of the opalescent squid die after spawning, and millions of their pale cadavers litter the ocean floor among the egg masses of the next generation.

The next day, equally frenzied activity takes place on shore. Dozens of trawlers have converged on tiny harbors to unload their catch. Normally the boats operate out of larger ports, but this annual glut of squid gives the fishermen a respite from the offshore fisheries. While the squid remain plentiful and the price remains high, the fishermen can make good money in this coastal fishery, putting off for another month the time when they must return offshore to their traditional fishing grounds.

In the last decade offshore fishermen have encountered new Japanese squid-jigging boats. At night their halogen lights illuminate the sky so brilliantly that the Coast Guard often receives reports of burning ships. The lights are used to attract vast schools of squid. The boats' fish masters scrutinize the ocean surface for reddish clouds of diatoms, which indicate the presence of squid. The squid feed on crustacea and small fish attracted to these delicacies.

Hundreds of lines are deployed over the side and elliptical drums raise and lower the metal jigs. This is a cruel ploy, for during the mating season males grasp the jigs in an amorous embrace only to be caught on a ring of sharp metal hooks. On a good night, 33,000 pounds of squid tumble over the side every hour.

Meanwhile, a quieter migration is taking place on land. Over fifty scientists and numerous students have converged on the village of Woods Hole. For the next four months, Nobel prizewinners, research assistants, and technicians will amble to the docks, buckets in hand, to retrieve their squid. By the end of the summer, they will have teased out fifteen thousand of the five-inch nerves that run down the length of the squid's mantle. They will have probed these nerves with electrodes, drugs, and optical microscopes. By autumn they may have elucidated a few more facts about the complex mechanisms underlying the nervous system, and they will have advanced a few steps closer to curing diseases of the nerves, such as Alzheimer's disease and amyotrophic lateral sclerosis (or Lou Gehrig's disease).

4

The Lobster Posture

... the Mock Turtle ... went on again:—

"You may not have lived much under the sea—" ("I haven't," said Alice)—"and perhaps you were never introduced to a lobster—" (Alice began to say "I once tasted—" but checked herself hastily, and said, "No, never")—"so you can have no idea what a delightful thing a Lobster-Quadrille is!"

"No, indeed," said Alice. "What sort of a dance is it?"

"Why," said the Gryphon, "you first form into a line along the sea-shore—"....

—Lewis Caroll, *Alice's Adventures in Wonderland*

It is night. The female horseshoe crab continues her long odyssey across the continental shelf toward the bay where she will lay her eggs. As she approaches the inlet to the bay, receptors along the top of her carapace sense the lower salinity and warmer temperature of the water. The muddy bottom and acres of waving eelgrass beds are now interspersed with rocks and boulders transported by glaciers twelve thousand years ago. As the crab crawls through one of these rockstrewn areas, she comes across a pair of mating lobsters.

The female lobster, sensing that her time to molt is imminent, has left her solitary burrow in search of a mate. For days she has patrolled the area to determine the location of the most dominant male. Now she stops at the mouth of his burrow and raises her body. An internal pump whirls into motion and a strong current of water flows through her gill chamber into the male's burrow. The water carries pheromones, chemical aphrodisiacs released from the urinary pores located just beneath her antennae.

The perfume is irresistible to the normally territorial male. He welcomes her into his burrow, where an elegant dance ensues. The male rises on the tips of his walking legs, arching his back off the substrate. Pleopods beneath his carapace pull a current of water containing the chemical signature of the female toward him.

The female has taken on the posture of an abject dog: head down, back arched toward the substrate, tail and claws extended. They reach out and touch with their antennae.

Now they pull back, boxing and jabbing at each other with outstretched, clashing claws. They push and pull like ungainly sumo wrestlers. Finally the female pulls back and turns away from the male. The gesture seems to appease his aggression. Jabbing his claws into the substrate, he quiets down. The initial courtship is over. They will now settle into an uneasy cohabitation.

As her time to molt approaches, the female becomes more restless. She "knights" the male by resting her claws on his rostrum. Pheromones engulf the male's chemoreceptors. The moment of molting is near.

The female lifts her third walking leg and suddenly stumbles over sideways. Her tail and legs thrash about while the male stands by, feeling her newly exposed body with the tips of his antennae. With a final kick, the female

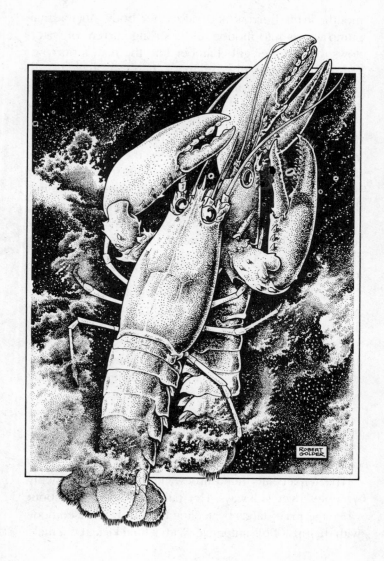

withdraws her tail from the old shell. The male waits for her recovery.

Now the mating time has arrived. The male gingerly mounts the soft female from behind. Supporting himself on his large claws, he gently turns the female on her back. She remains limp, abdomen stretched and claws extended. Their pleopods are beating rapidly.

The male slowly moves forward placing his gonopods firmly into the females waiting annulus. With two or three thrusts, a spermatophore is deposited into the female's sperm receptacle, where it will remain for twelve months. The male releases his hold and dismounts, the female's tail flips from beneath his body, and she retreats to the corner while he quietly eats her old shell.

It is the female's shell that has led to some extraordinary evolutionary compromises. Like most cold-water arthropods, the lobster molts only once a year. It is only after molting that the female can mate, and thus evolved a ritual in which she actively seeks out a dominant male, to win his protection while she is without a shell. Both benefit from this ceremony: the female gets protection, the male gets sexual exclusivity by preventing her from mating with rivals.

Thus, there is a strong incentive for the evolution of a courtship ritual that makes the female selection of a mate possible. This has induced evolution to work in ingenious ways with several neurological systems.

Two important mechanisms underlie the courtship behavior. Lobsters' bodies are covered with millions of fine, hair-like receptors that make up a chemoreceptor system a million times more sensitive than man's simple sense of taste, or smell. Normally this elaborate system is used to detect distant food and the detectors are specific to individual amino acids, the building blocks of the proteins found in animal tissue. The information about different amino acids is transmitted to the lobster's pea-sized brain, which assembles a mosaic of chemical signals representing the signature of the other animal, whether a mate, enemy, or food. The female relies on the male's sophisticated taste system to inform him of her intentions.

Dr. Ed Kravitz, a neurobiologist at Harvard Medical School, has studied the role of two amines that act as neurohormones in regulating the lobster's dance. The amines serotonin and octopamine are released from nerves near the brain into the blood system. The blood carries the neurohormones to exoskeletal muscles where they regulate stereotypical postures. In the male, serotonin causes all the flexor muscles to contract, creating his dominant posture, while in the female, octopamine causes all the extensor muscles to contract, thus inducing her submissive posture.

Essentially, communication between the lobsters is similar to the communication between nerves and muscles. Pheromones and visual postures trigger the release of neurohormones, activating exoskeletal muscles which cause stereotypical postures and behaviors. The three components—pheromones, visual stimuli, and neurohormones—must all interact according to a set pattern. Each posture must be answered by the corresponding posture in order for the courtship to proceed.

In the end, if all goes according to plan, mating occurs. The female has benefited from insemination and protection during her time of maximum vulnerability and the male has benefited by passing on his genes while preventing the willing female from mating with his rivals.

The female horseshoe crab, not a voyeur by nature, proceeds on her migration.

5

At an Ancient Orgy

It is June. The moon is full. There are no sounds save for the quiet lapping of the bay against the shore. The sand-flats await the flooding of the high-course tide, the highest of the month. An expectant sense of creation fills the heavy night air, for it was in warm shallow waters like these that life first arose.

The female horseshoe crab has waited for this convergence of events in order to repeat a ritual that has persisted for eons. Now she lies in ten feet of water bathed in the silvery luminescence of the full moon.

She is not alone. A smaller male crab crawls toward her and executes a brief circling dance before clasping her carapace with his specially modified mating claws. Tonight his tenacity will pay off.

As they make their way toward shore, the couple must

navigate through a stag line of eager male suitors. By the time they reach the shore, the female crab will be surrounded by an entourage of thirty or forty lascivious male crabs.

The tide has now reached its peak. The invisible forces of the sun and the moon draw rivulets of shallow waters up the beach. The watery phalanx advances across the sandflat, pauses at the high-tide mark and slowly starts to retreat. It is the furthest incursion the bay will make upon the shore until the next spring tide thirty days later.

Thousands of horseshoe crabs have followed the advancing tide up the beach. The dark forms of their shells are silhouetted against the sheen of the moonlit sandflats. All that can be heard is the quiet scrapping and scratching of their shells as the ungainly creatures clamber over each other in their eagerness to lay and fertilize their eggs.

The female digs a few inches into the moist sand just below the high-tide mark. Her strong sand-pushing legs propel her deeper into the substrate until she almost disappears. After resting she starts to deposit her eggs. Hundreds are released before she drags her consort male over the spot so that he may fertilize them. The rest of the males crowd about her, eagerly competing to fertilize any eggs the primary male might miss.

Their time is short. The tide is fast receding and the deadly heat of the morning sun will soon be on them. One by one, the males disperse and crawl back into the quiet waters of the bay now awash with eggs, sperm, and pheromones.

Within thirty minutes it is over. All that remains below the sand where the female dug is a small bolus of thousands of tiny green eggs. They will stay buried until the waters return on the little spring tide fourteen days later.

As they have for 500 million years, the alignment of the moon, the sun, and the earth have triggered the same spring mating ritual on opposite sides of the earth. Formerly a single species of horseshoe crab enjoyed worldwide distribution along the temperate water coasts of ancient continents long since disappeared.

Today four species of horseshoe crab remain, three living along the east coast of the Orient and one, *Limulus polyphemus,* along the east coast of the United States. It is the warm waters of the Atlantic Gulf Stream and the Pacific Japan Current that make the east coasts of these two areas habitable to modern horseshoe crabs. Presumably, similar currents generated by the gyre of the Tethys Sea made the east coast of Gondwana Land habitable to ancestoral horseshoe crabs millions of years ago.

Limulus polyphemus, a confirmed East Coast dweller, ranges from Maine to the Yucatan. The center of its population is in Delaware Bay. Here the adult females, always larger than the males, might be two feet long and weigh up to three pounds. During the mating season hundreds of thousands of the crabs swarm over the beaches in an impressive orgy of fecundity. To the north and the south, both the size of the populations and of the individuals diminish because of less favorable temperatures.

Horseshoe crabs remain much as they were millions of years ago. Closely related to modern spiders, horseshoe crabs are not true crabs but are members of an ancient order of animals called Chelicerata because of the two tiny feeding claws or chelicerae that push food toward their mouths. In addition to these, the horseshoe crab has four sets of legs intricately adapted to meet its needs.

When the males reach their third year, a form of adolescence occurs. The forward pair of true legs change during molting to become a fist-shaped structure admirably suited for clasping the female during mating.

The crabs also have a long spiky tail or telson, fearsome in appearance but benign in reality. Often assumed to be a weapon, the tail is actually used as a lever to turn the crab over if it ends up on its back.

The crab's mouth lacks jaws and is located amidst this impressive collection of legs. Bristles at the base of the legs masticate the crab's food so that it must be walking in order to eat effectively. Behind the legs is a large set of book gills similar to those found in spiders. These sometimes double as swimming appendages. The crabs use them occasionally to perform a clumsy kind of upside-down backstroke.

But the most arresting feature of the horseshoe crab is its eyes: a pair of compound eyes facing the side; a pair of simple eyes sensitive to ultraviolet light facing forward; and five vestigial eyes situated below the carapace. Surely this curious array of eyes must serve some important purpose. Some scientists speculate that these vestigial eyes originally evolved when horseshoe crabs spent more time swimming on their backs.

Indeed much of what we know about nerves and vision we know from research on horseshoe crab eyes. But we must save a discussion of that for a later chapter. For now let us leave our female horseshoe crab awash in the life sustaining waters of the bay resting from her exertions at the horseshoe crab orgy.

PART TWO

SUMMER

Marine Animals
and Modern Medicine

6

~~~~~~

Birth and the
Genetic Sex Switch

The spring tide has turned. The orgy is over. One by one, each spent horseshoe crab returns to the bay. Now only the predators remain. Minnows, eels, and crabs patrol the receding waters, gorging on the eggs that have washed out of the sand. Hundreds of shorebirds probe the wrack line to feed on this annual bounty.

The abundance of this readily available protein has affected the migratory behavior of at least one species of bird. The red knot, *Calidris canutus rufa,* is a small shorebird that rivals the better known arctic tern in the length of its migrations. It nests north of Hudson Bay on the Canadian arctic tundra and migrates south to Tierra del Fuego on the southernmost tip of South America.

During its migrations it stops at specific locations where it can feed on sure and abundant supplies off protein-rich food. One of these locations is Delaware Bay, where one hundred thousand red knots (roughly one half the world population of these birds), congregate to feed on horseshoe crab eggs. During their brief feast they gain more than forty percent of their body weight. The protein reserve gained from the horseshoe crab eggs fuels their flight and is crucial to the female's egg-laying shortly after reaching the nesting sites.

Meanwhile, only a few inches below the surface of the moist sand, the pinhead-sized green eggs of the horseshoe crab are preparing to undergo their own significant change, one which recapitulates their evolution and unites all higher forms of life. It is the much studied but little understood development of the embryo.

Within minutes of insemination, the single cell of each egg starts to rotate. Yolk flows back and forth and fissures form. The cell begins to divide. Granules appear and disappear on the surface, giving the merest hint of the profound changes occurring within. It is a preprogrammed sequence of events that will continue unabated until birth.

By the fifth day rudimentary appendages have formed. On the sixth day the egg is ready to undergo its first molt, one of the horseshoe crab's hallmarks as an arthropod. The tiny embryo starts taking in water and swelling. A rent appears in the thin membrane that surrounds it and with spasmodic jerks, the tiny, ill-formed embryo extricates itself from the clear exuviae. Yet the embryo has not hatched—it still rests within an egg of two shells.

On the seventh day the embryo swells again. It sheds its membrane as before but this time the swelling continues. A rent appears in the chlorion, the green outer shell of the egg. Suddenly the chlorion breaks and the egg swells in size to twice its former diameter, hatching into what is called the transparent bubble egg. It will molt twice more before this membrane will also break, releasing tiny trilobite larvae into the life-sustaining waters of the bay.

The embryo has developed in synchrony with the movements of the heavens. On the fourteenth day the earth, sun, and moon will be in line, producing the little spring tide, whose waters flood the nest and help break open the last of the unopened eggs.

Why does the horseshoe crab go through this elaborate procedure? Why did it, a marine animal, start laying eggs on land? The evolutionary rule of thumb is that animals are conservative when it comes to laying their eggs. They return to the environment for which they were originally suited: Thus, reptiles such as turtles return to land and amphibians such as frogs return to water when laying their eggs.

There is some fossil evidence that relatives of the horseshoe crab were once freshwater animals that fed on insects. But is it possible that something else is going on? If the horseshoe crab was around before there were reptiles, birds, and terrestrial mammals, it would have been an excellent strategy to lay its eggs on land, where they would be out of the reach of marine predators. Even now when horseshoe crabs lay their eggs, they are surrounded by minnows, eels, and catfish, which swim in to devour the succulent morsels. As the spring tide recedes, the eggs are safe from these aquatic predators.

What is the purpose of the double shell? Horseshoe crabs face a difficult problem: their eggs are laid between high and low tides. The eggs must contend with the dramatic extremes of this environment. If it rains before the tide advances, the salinity will plummet; if the sun shines, the temperature will rise. When the tide comes in the situation is reversed: the salinity rises and the temperature plummets.

The double shell protects them from these extremes. Safe within the inner shell, the embryo can alter the salinity of its environment through osmosis. It is almost a precursor to the mammalian womb.

But the complications go further. The eggs laid on the upper beach grow faster because the sun warms their nest, speeding up their growth. Some eggs are laid too close to the water; if the water temperature drops too low, it will arrest their development. They can remain in this arrested state for up to eight months. But as soon as the temperature rises above 68 degrees Fahrenheit, the process will resume where it left off.

Perhaps there is another phenomenon occurring here. Scientists have recently discovered that in many animals the temperature of the surroundings determines whether males or females will be born. In one experiment they found that if turtle eggs were moved to the upper beach, they would all hatch out as males. If moved closer to the water, they would all hatch out as females. This also proved to be true for larger reptiles such as alligators, which lay their eggs in simple nests like those of the early dinosaurs. Curiously, more males are born when the temperature is either higher or lower than normal.

Is it possible that there is a genetic switch that determines sex? Is this switch activated when the temperature is too hot or too cold? If that is the case, there may be a good reason for it. According to the dictates of sociobiology, animals have strategies to improve what is known as their reproductive fitness.

One strategy for improving the chances that offspring will be reproductively successful is to control their sex. Females can only produce a certain number of eggs, but males can fertilize infinitely more. This usually means that females will adopt a "conservative" strategy of laying fewer eggs, but put more care into raising them. Males, on the other hand, may adopt a "radical" strategy by fertilizing many eggs on the chance that some will do better than others. If times are good and climatic conditions favorable, it makes more sense to adopt a conservative strategy. If times are bad or the climate is changing, it might be better to adopt the "radical" strategy.

Under normal conditions a female might have a better chance of producing more offspring. So, for instance, if a horseshoe crab is mating near the center of her habitat range, where the temperature for hatching eggs should be favorable, it would be better to adopt a "conservative" strategy by having more female offspring.

If an animal happens to be on the periphery of its range, however, where environmental conditions may be less favorable, it might make more sense to adopt a "radical" strategy by producing more male offspring. A male can fertilize hundreds of thousands of eggs. If the season is mild, he might be able to father hundreds of thousands of offspring while a female could be responsible for only a few thousand. To put it another way, if an egg finds itself on the edge of its range, it would be a better evolutionary bet for it to become a male. If it finds itself in the middle of its range, it would be better to become a female. Thus we might expect the egg to have a genetic sex switch:

turn to male if it's too hot, stay female if it's just right.

There is some intriguing evidence that a similar process occurs in mammals. During bad seasons, red deer will produce more males than females. In baboons it has been observed that high-ranking females tend to have female offspring, a conservative strategy to maintain their reproductive fitness. Low-ranking females tend to have males.

Perhaps we should think of males as part of a huge evolutionary experiment. There were no males until there was sex. Contrary to the myth of Adam and Eve, males arose from females. Because one male can mate with many females, they are highly expendable. They can thus be used by nature as experiments to try out new evolutionary strategies such as expanding one's range or adopting a new migratory behavior.

As we learn more about the intricate workings of the genes, we are discovering an incredible ability to affect behavior. A genetic sex switch triggered by temperature might be one more way to improve the reproductive fitness of animals. If this is true, then is it possible that the same thing occurred among the dinosaurs? If volcanoes spewed tons of ash into the atmosphere or if a meterorite crashed into the earth, did the resultant cooling of the climate foil the genetic sex switch? Did the dinosaurs die because there were too many males?

7

The Thorn in the Star, The Gout in the Fish

Creeping slowly across the floor of the summer estuary, a five-armed starfish senses the presence of a feeding scallop. Chemical substances spewing from between the scallop's gaping shells tell the starfish that its quarry is near. One long arm draws close to the feeding mollusk.

Suddenly the scallop snaps its shells shut and skips up off the bottom. A burst of water forced from between the shells jet-propels the scallop into the water column. A few more squirts carry it safely away from its enemy. The mollusk drifts quietly back to the ocean floor, only to land on the waiting arm of another starfish.

This starfish is more fortunate. It reaches up and envelops the scallop in the deadly embrace of its five strong arms. Hundreds of tiny suction cups clasp onto the scallop's shell. Aided by a highly developed water vascular system, the starfish starts the long procedure of pulling the scallop apart.

The constant pressure of the starfish arms gradually overcomes the scallop. Each time the mollusk, relaxes its abductor muscles, the starfish gains a slight advantage. Millimeter by millimeter, the shells are pried apart. After several hours of this exhausting struggle, the scallop is dangerously weak and can no longer prevent the starfish from everting its slimy stomach through its mouth and down into the scallop's body cavity. As the starfish extrudes its stomach, strong digestive juices flow over the mollusk's soft body tissues. The process of digestion begins. After a few more hours, the grisly meal is over. The starfish glides away.

Overhead, a pack of spiny dogfish, *Squalus acanthias,* partake in a far more violent meal. A feeding frenzy erupts as thousands of the small gray sharks slash and tear into a school of herring. Sharp teeth grasp and flay at quivering flesh. Heads shake and dull-gray abrasive bodies twist with the exertion.

One of the female dogfish breaks off from the pack and swims toward the bottom, her stomach bulging with her young. Two pups have been developing within her uterus for almost two years, one of the longest gestation periods of any animal. One by one the two pups will wriggle out of their mother's womb and, within an hour, they too will be tearing great hunks of flesh out of herring twice their size.

This prolonged gestation gives the dogfish a significant advantage over most fish, which simply broadcast millions of eggs into the ocean to contend with the vagaries of nature. The spiny dogfish's birth process ranks as an intermediate stage, falling between the egg-spawning fish and live-bearing mammals. Although the dogfish's pups developed inside her uterus, each was encased in a leathery pouch, called a candle, and was nourished by a protein-rich yolk. Other sharks have taken the evolution of live bearing one step further.

For example, some female sharks secrete a type of uterine milk. Others have developed a gruesome system of nourishment befitting their vicious personality: the firstborn and larger pups eat their brothers and sisters while still in the uterus. By the time of birth, only a few large, well-fed pups are born.

Curiously, starfish and dogfish have come together in the laboratory in a series of experiments which spanned a century, and helped unravel one of the basic mechanisms of human disease.

One of the first and most venerable example of how marine animals have been used to understand human disease occurred in the humble apartment of Élie Metchnikoff, a then destitute biologist who would become known as the father of biomedicine.

In the winter of 1883, the Russian-born Metchnikoff was sitting in his living room overlooking the Straits of Messina, in Sicily. He had just been fired from a university position and had converted his living room into a small lab. His family had gone into town to see some performing apes at the circus. He was going to use the time to observe the moving cells of a starfish larva under his microscope. As he watched the cells, Metchnikoff was suddenly seized by a revolutionary concept: perhaps the motile cells could act as the body's main defense against foreign invaders. He paced up and down the room, and then hurried to the sea to collect his thoughts.

Until that time doctors believed that blood vessels, the nervous system, or changes in body temperature were somehow responsible for fighting infection. The transparent starfish larvae had neither blood vessels, a nervous system, nor the ability to regulate its temperature. Without these advanced attributes, Metchnikoff surmised that it had to be the mobile cells themselves that were responsible for attacking foreign invaders.

Metchnikoff raced into his garden where the impoverished family had converted a living tangerine tree into a makeshift Christmas tree. He plucked a thorn off a rose bush and inserted it under the skin of the starfish larva.

The next day, after a fitful sleep, Metchnikoff went straight to his microscope. Even in the feeble early morning light, he was able to verify that his experiment was a success. Cells had migrated to the area of the intruding thorn and clumped around it.

That simple experiment was the beginning of our understanding of the role of cells in fighting human disease. Virtually all our modern medical advances owe their origins to that basic understanding and the microscopic perspective used so effectively by Metchnikoff.

After that simple experiment, Metchnikoff spent the next twenty-five years working on phagocytosis, his theory of how cells fight infection by consuming invading organisms. But it is interesting to note how the behavior of cells attacking infection is so close to how starfish attack mollusks. Undoubtedly it was during Metchnikoff's fitful sleep that the two images merged, sparking one of mankind's most creative flashes of insight about how our bodies work.

After that discovery, Metchnikoff's prospects improved greatly. He went on to positions at the Bacteriological Institute in Odessa and at the Pasteur Institute in Paris, and he shared the Nobel prize in 1908. More importantly, his insight was to change medicine forever.

Up until that time medicine was a rather hit or miss affair. Doctors with little more sophistication than medicine men had built up a slender pharmacopeia of substances that seemed to help fight certain diseases, but they didn't know why these drugs were effective. In fact, many of our most effective drugs, such as morphine, quinine, and digitalis, were plant extracts first discovered by witch doctors. Today we use these drugs to help relieve pain, cure malaria, and prevent heart attacks.

If plants had evolved such powerful substances to protect themselves against insects and other animals that might graze on their leaves, then certainly marine organisms through their millions of years of evolution should have developed chemicals to protect themselves.

The watery environment of the ocean is ideal for the transport of chemical substances like those emitted from scallops and starfish. By trial-and-error testing of the billions of pheromones, hormones, toxins, and neurotransmitters produced by marine animals, modern scientists, like witch doctors, should be able to come up with at least a few more drugs for medicine.

But this is as arduous a task as it sounds. It is like looking for the proverbial needle in a haystack. Wouldn't it be easier if you had a tool, perhaps a magnet, for separating the needle from the hay? The magnet of biomedicine is the use of animals as models for understanding human disease.

THE YEAR OF THE CRAB

Once scientists know that a specific bacteria like strep-tococcus causes scarlet fever, they must isolate an anti-biotic like penicillin or streptomycin to help cells kill those invading bacteria. It was basic research on many infec-tious diseases done by Metchnikoff and others in the late 1800s that led to the drugs that conquered these diseases in the 1930s.

Today it is basic research on diseases like cancer, arte-riosclerosis, arthritis, schizophrenia, and acquired immune deficiency syndrome (AIDS) that will allow scientists to discover direct cures. This will end our present reliance on costly methods, such as surgery and chemotherapy, which only treat the diseases indirectly.

The advantages of using relatively simple primitive marine animals, such as horseshoe crabs and starfish, are spectacular. Because they remain examples of early stages of planetary life, they retain simple mechanisms to fight disease evolved before the appearance of more complex backup systems.

A modern example of how marine animals can be used to understand and cure diseases involves the smooth dogfish. Again, modern scientists stood on the shoulders of Metchnikoff to understand the mechanisms of one of mankind's oldest scourges: gout.

Gout is a type of arthritis, that particularly affects the joints of the big toe. It has long been associated with diet. Foods that were known to contribute to gout include: gamey foods, like ducks, geese, and venison; organ foods, like kidneys and liver; and wines, which are high in pu-rines, a form of uric acid. That delectable food list leaves little question as to why gout was long known as the "rich man's disease."

Since the early 1960s doctors have suspected that gout was caused by crystals of monosodium urate formed by the metabolism of purines. When the concentration of monosodium urate was too high in the blood system, crystals would precipitate and collect in the joints. Here the crystals were attacked and ingested by phagocyte cells, just as Metchnikoff had observed in starfish larvae. Only with gout, about thirty minutes after the crystals were ingested, the white blood cells died. Scientists felt that if they could observe this mechanism in living cells, they could not only discover the molecular basis for gout but understand the underlying basis for inflammation, perhaps the most fundamental of all disease processes.

But to do this, they would have to use non-human material because the critical organelles of human white blood cells are too small, barely visible under a light microscope. They turned to another shark, the smooth dogfish, *Mustelus canis,* whose cells and organelles were easily visible and could be filmed under a microscope.

In 1975 Doctors Gerald Weissman and Sylvia Hoffstein, working at the Marine Biological Laboratory in Woods Hole, added crystals of monosodium urate to a culture of dogfish blood. Like the cells in Metchnikoff's starfish, the white cells of the dogfish migrated toward the crystals and ingested them.

Then a curious thing happened. The cells proceeded to commit suicide, much like a spy swallowing his hidden cyanide pill. As needle sharp crystals were ingested, they pierced a sac of enzymes, poisonous to the cells. When the crystals pierced the sac, the enzymes quickly killed the cell and flowed out into the joints, where they caused acute inflammation. The crystals, largely unaffected by the procedure, were then ingested by other living cells and the pattern was repeated.

Today, thanks to starfish, dogfish, Metchnikoff, Weissman, Hoffstein, and hundreds of other workers, few people suffer the debilitating effects of gout. Once the disease was understood, its prevention and cure were made possible.

Now an impressive array of drugs allows clinicians to prevent gout by controlling the formation of monosodium crystals in the blood. An equally impressive collection of anti-inflammatory drugs is allowing doctors to ease the pain of other, less controllable forms of arthritis. Someday soon, we can expect that researchers building on the work of Metchnikoff and Weissman will unravel the mechanisms of these other crippling arthritic diseases.

8

Ecdysis:
Horseshoe Crab Adolescence

It is early summer. The shallow waters of the bay are replete with tiny swimming larvae, the planktonic forms of billions of bottom-living animals. The early Cambrian seas must have looked much like this long before the evolution of fish. Motes of life, these larvae are the future clams, crabs, and worms of the estuary. They will enjoy this brief swimming existence for a fortnight before settling to the ocean floor to take up their sedentary adult life.

Swimming erratically through this teeming soup of living animals are thousands of tiny horseshoe crabs. Called trilobite larvae because they look like the ancient ancestors of horseshoe crabs, they are a zoological curiosity that has stirred naturalists' imaginations for over a century. The tiny forms have a yolk and will swim for several days before undergoing another profound change: ecdysis, or the molting of the hard chitinous shell signifying the coming of age, the adolescence, of horseshoe crabs.

One thing the reader must understand about marine biologists is that they are generally more curious about sex than other members of society. Ed Ricketts, friend, mentor, and disguised protagonist in many of John Steinbeck's novels, is perhaps best remembered for the interest and ministrations that he bestowed on his neighbors in the house of ill-repute next to his laboratory on Cannery Row.

There is strong circumstantial evidence that an enterprising striptease artist in Baltimore also counted an erudite marine biologist as one of her guests. She billed herself as the world's foremost ecdysist artist. Ecdysis is the Greek-rooted word that marine biologists use to describe molting. It pertains to the shedding of outer garments: feathers in the case of the stripper and a hard chitinous shell in the case of the horseshoe crab. Not to deny the stripper her rightful place in history, I would like to point out the horseshoe crab's longstanding familiarity with the process.

Scientists have spent considerable time and effort trying to unravel the intricacies of the horseshoe crab's ecdysis. They have irradiated horseshoe crabs' rudimentary eyes with lasers, injected their blood with plant and animal extracts, and tied off parts of their bodies in an unsuccessful attempt to understand the exact nature of the process.

Researchers believe the hormones that regulate ecdysis

in horseshoe crabs come from glands near the eyes, as they do in lobsters. One way to make a lobster shed its shell and thus grow faster is to cut out its eye, a radical solution at best.

The newly hatched trilobite larva molts at least six times in its first year and several times in its second year. As it grows older it sheds fewer times and takes longer in the process. Quite often people find the discarded shells on the beach and think they have found a dead crab. A closer look will reveal a narrow crack along the leading edge of the crab's shell. It was through this crack that the crab crawled while extricating itself from the old shell.

After a few molts something new and insidious may appear on the crab's gills: tiny black spots, which indicate that a small flatworm, *Bdelloura candida,* has laid its eggs in the gills of the immature crab. The crab is now in danger. As the flatworm eggs develop they can tear the tissue of the crab's gills. This will allow bacteria to enter the horseshoe crab's bloodstream; the results can be fatal. If the crab is not able to kill off these bacteria, it will die.

Unfortunately these bacteria are as ubiquitous as they are fatal. They thrive in warm shallow waters, as they have since the Cambrian. Undoubtedly the horseshoe crab's ancient ancestors had to contend with similar flatworms and bacteria. It is fortunate that they did because they have preserved one of nature's most effective antibiotic systems, one which mankind has started to use in its own unending struggle against bacterial disease.

It was in the Paleozoic crucible that horseshoe crabs first evolved the primitive but elegant mechanism for fighting bacteria. The discovery of how horseshoe crabs fight bacteria is one of the most fascinating stories of modern research. But to tell the story fully we first must follow the female horseshoe crab back to Woods Hole, Massachusetts.

9

Penikese and Agassiz: Woods Hole Beginnings

It is mid-July. The sun beats down on the quiet waters of the bay and life proceeds with its slow and elegant summer pace. A male horseshoe crab still clasps the female's carapace, though it has been over a month since she last laid her eggs. Now they crawl across the substrate feeding on worms, crustacea, and mollusks.

The tide turns and deep within the sediments the pair sense the change. They push their way to the surface. The sun shines on the sandy floor and ribbons of sparkling green eelgrass sway in the incoming currents.

Overhead a shallow draft boat is being poled through the shoal waters. In the bow a sharp-eyed man stands with net in hand. "To the right, to the right, starboard, starboard, there's a pair!"

A shadow crosses the crabs' horny-plated eyes. Sensitive to polarized light and able to magnify sunlight tenfold, their powers prove to no avail. A hand descends through the surface and grasps the male crab by its tail. The hapless pair are unceremoniously pitched into the bottom of the boat. Still attached, they land upside down, legs flaying in the air.

The harsh sunlight pulses down on their vestigial ventral eyes and their heart rate races, then plummets radically. They lie on several hundred brethern horseshoe crabs about to embark on an odyssey through modern biomedicine. Their destination: the Marine Biological Laboratory (MBL) in Woods Hole, Massachusetts.

Horseshoe crabs, Woods Hole, and the MBL have been associated with each other for over a century. But it all started on Penikese, a small island fifteen miles off the coast of Woods Hole. In 1873 Louis Agassiz, an imposing Harvard professor, established the Anderson School of Natural History, a small summer laboratory on Penikese Island.

At the opening of the laboratory, the imposing patriarch, Dr. Agassiz, intoned to his assembled students the creed of his lab: "The grand purpose of this school is the study of nature. I will try to make you investigators, to teach you to find out what you want to know for yourselves, that you may be able to do the same thing in other places where you may have no guide. But there is one thing about which I am certain, that we not begin on our task by reading books."

Today a shortened form of his dictum—"Study nature not books"—hangs, appropriately, in the MBL library, the largest marine library in the world. I am sure it would please the old teacher to know that his words still send quivers of guilt through thousands of the world's most emminent biologists every summer at the MBL.

Over one hundred years later, the Anderson School's child, the MBL, flourishes. During the winter it houses a few score of hardy biologists wily enough to have gar-nered funding from the government. But every summer, however, it blooms with vitality. Over a thousand gradu-ate students, full professors, and Nobel laureates descend on the laboratory to study nature. They talk, argue, think, and sometimes even read in the process. By autumn they return to their respective universities to publish the results. It is not an understatement to say that their summer experiences have played a major role in the startling advances made in biology and medicine during the last half century.

I like to think it has also affected literature, for Ger-trude Stein did some time here. Could that account for a certain scientific objectivity to her writing, to wit: "Rose is a rose is a rose is a rose"?

But the MBL was not an isolated phenomenon. Prior to the turn of the century marine laboratories sprang up in Italy, France, and England partly in response to Darwin's theory of evolution. Scientists thought that the key to understanding the underlying mechanism of heredity would be found by studying embryology. They were wrong, but we still remember Ernst Haeckel's statement: "Ontogeny is the short and rapid recapitulation of phy-logeny determined by the physiological functions of heredity (propagation) and adaptation (nourishment)." Or, as it is known to many biology students, "Ontogeny recapitulates phylogeny."

Marine animals have recently proved ideal models for

the study of human biological systems and disease. Because marine animals tend to be more primitive and simple than terrestrial forms, their cellular structures and embryological development are easier to understand. That understanding has led to the almost unimaginably rapid advance in our knowledge of the secrets of life and disease.

From the beginning the MBL has had an able collecting department that scours the waters surrounding Cape Cod for interesting creatures to study. Of the two hundred fifty species readily available to the researchers, the horseshoe crab and the Woods Hole squid have proved most fruitful, while lobsters have proved most flavorful. Scientists requesting them are often suspected of popping them into boiling beakers at the conclusion of their experiments.

Today the MBL continues in its falteringly efficient manner. Like the founders of the United States, the original founders of the laboratory distrusted central power. This comes easily to anyone familiar with the workings of evolution. They established the institution so that it would be owned and operated by the scientists who use it. Like evolution, this seemingly inefficient system works. All the component parts collide and stumble over each other in an haphazard attempt to arrive at a better way to survive. It might be a good thing to keep in mind as we take up the techniques of gene splicing to shape our own evolution.

10

Serendipity:
The Discovery

Scientists, like artists, musicians, and mythical princes, are occasionally smiled on by the gods. It happens infrequently, but its consequences have been so profound that the phenomenon has an official term: serendipity. Serendip, as noted in the Koran, was the island state off India where Adam fled after being exiled from the Garden of Eden. We now know the nation as Ceylon or Sri Lanka.

It is fabled that three princes from Serendip set out on a journey to discover treasure. They never found it, but they returned with something far more valuable: knowledge unsought, but happened upon accidentally.

Serendipity has played an important role in medicine's quest to understand disease, but it is often misunderstood. Serendipity is not a form of blind luck. Rather, it is a form of inspired luck that originates from asking the wrong questions. It is only the truly perceptive scientist who recognizes that the wrong answer to his question is really the correct answer to a more insightful question.

Serendipity has played a profound role in our centuries long quest to conquer bacterial infection. In 1928 Sir Alexander Fleming noticed that one of his bacterial cultures had become infected by a common green mold usually found on stale bread. Another scientist might have thrown out the culture and started over. But, intrigued by the ring surrounding the mold where no bacteria was able to grow, Fleming saved the culture and began experimenting with it. Eventually it was realized that the mold, *Penicillium notatum,* inhibited the growth of some forms of bacteria, and thus the first antibiotic, penicillin, was discovered. However, penicillin only inhibited certain types of bacteria. Since the 1880s scientists have recognized two major families of bacteria. They are called gram-negative and gram-positive after Dr. Hans Christian Gram, who discovered the staining technique still used to distinguish between the two groups.

Later, when scientists were developing a vaccine for typhoid fever, they discovered that the cell walls of heat-killed gram-negative bacteria caused the burning fever associated with diseases caused by gram-negative bacteria. They used the term pyrogens—"burning bodies"—to describe the virulent endotoxins produced by the polysaccharide cell walls.

As ubiquitous as they are lethal, gram-negative bacteria have existed for hundreds of millions of years and are found today in almost all watery environments, including our intestines, where they do little harm. If the endotoxins enter the blood system, however, the result is often fatal. This is what happens after an accident when a person goes into shock. Trauma breaks down the membrane between the intestines and our blood system, allowing endotoxins to invade the bloodstream. A high, often fatal, fever develops.

Dealing with pyrogens has become an urgent problem. Within the past few years, scientists have discovered several virulent forms of gram-negative bacteria that are resistant to antibiotics. Ironically, these bacteria are most pervasive in modern hospitals. Colonial Americans never faced this problem because hospitals were routinely burned after a few years of use—a measure frowned on in today's world of rising hospital costs.

A simpler method would be to devise a system to check for pyrogens. In fact, this must be done every time an instrument is used in surgery or an intravenous needle is inserted into a patient's bloodstream. It also must be done when a patient is admitted to a hospital with a suspected case of a gram-negative disease, such as toxic-shock syndrome or spinal meningitis. Blood, urine, sputum, and spinal fluid samples are rushed to the lab. The gram stain is then used to identify if gram-negative bacteria are present. If gram-negative bacteria are present, the samples can then be injected into a live rabbit. Should the rabbit develop a severe fever, the bacteria is identified as a meningococcus, a genus of bacteria that attacks the meninges (the three membranes enveloping the brain and spinal cord at the back of the neck). With this diagnosis, the patient can be injected with a serum of rabbit blood containing specific antibodies to fight the bacteria of spinal meningitis.

Unfortunately, it takes up to forty-eight hours to obtain

the correct diagnosis using this method. With most gram-negative diseases, that delay can be fatal. A more effective method would be to devise a more rapid test for pyrogens. It would have made sense many years ago to investigate a primitive invertebrate, such as the horseshoe crab, to learn how it fights the gram-negative bacteria that are almost as common in its shallow-water habitat as in our own intestines. Of course that would have been asking the correct question. This happens less in research than the Nobel prize board would have us believe. Instead that old ally, serendipity, turned up in the MBL laboratory of Dr. Frederick B. Bang.

Dr. Bang was studying the circulation of blood in horseshoe crabs when he happened to notice that one of his crabs sludged up and died in response to a Vibrio bacteria. Most scientists would have discarded the crab and continued their work with another. Luckily Dr. Bang recognized that something unusual was occurring.

On investigation he found that horseshoe crabs possess large amoebocyte cells. When a horseshoe crab receives a wound, these primitive cells swarm to the area, coagulate, and kill the invading gram-negative bacteria. Most of the time this protects the crab by killing the bacteria. Occasionally, however, it protects the bacteria by killing the crab. Dr. Bang recognized that this reaction was strikingly similar to how rabbits respond to gram-negative bacteria. It is also an example of one of the most subversive ideas in medicine: that disease can result from the normal functioning of a body's defense system.

Dr. Bang immediately went to the head of Johns Hopkins University Medical School to suggest that a student be allowed to work on the phenomenon. The dean insisted hat his medical students were far too busy to spend time on horseshoe crabs. Dr. Bang cleverly bided his time until summer, knowing that most administrators think more clearly on vacation.

When he heard that the dean would be stopping in Woods Hole to catch the Martha's Vineyard ferry, Dr. Bang innocently invited him to visit the MBL. When the dean arrived Dr. Bang had a horseshoe crab ready. On seeing the clotting process firsthand, the dean demurred. By autumn, Jack Levin, a rising young hematologist, joined Dr. Bang's laboratory.

Their combined efforts led to the 1964 discovery that horseshoe crab amoebocyte cells could be used as an exquisitely sensitive diagnostic test for pyrogens. The stage was set for the commercialization of horseshoe crab blood.

11

Gold Extracted
from Blue Blood

The female crab is cold, stiff, and traumatized. She spent the night lying on the bottom of the collecting boat under several hundred other crabs. In the morning she was transferred to a large refrigerator truck and driven to the main laboratory of the bleeding company.

The truck is opened. Fifty crabs at a time are thrown into gray trash barrels and carried to the laboratory where rows of plastic bleeding racks stretch along the walls. The female crab is bent double and forced into one of the racks. Only the soft membrane above her heart is exposed.

A large stiff needle, the kind that veterinarians use for injecting horses, is inserted through the membrane into the pericardial sac surrounding the heart. Bluish-gray blood spurts out of the puncture and flows down the needle into a hundred-milliliter flask. On meeting the air, the copper-based blood oxygenates, frothing and foaming into a cobalt blue supernatant.

Horseshoe crabs have become big business along the East Coast. Their processed blood, called *Limulus* amoebocyte Lysate, is presently worth up to $15,000 a quart. Pound for pound, horseshoe crabs have become our most valuable marine animal, commanding prices once reserved for lobsters and salmon.

The story of the commercial development of *Limulus* lysate shares the intrigue of the gene-splicing and computer industries during their pioneer days. The road from basic research to commercialization is fraught with risks, competition, and money.

In this instance the "soul of the new machine" arose in the singularly pragmatic mind of Dr. Stanley Watson, a microbiologist at the Woods Hole Oceanographic Institution. In the early 1970s, Dr. Watson established a small real estate company called Associates of Cape Cod. At about the same time he started to use *Limulus* lysate to test the purity of membrane components from a marine bacterium. Gradually Dr. Watson realized that he could make a lot more money in horseshoe crabs than condominiums. He went to the Woods Hole Oceanographic Institution and the MBL for financing. The institutions turned him down, as did several pharmaceutical firms.

Undaunted, he convinced the Woods Hole Oceanographic Institute to sell him the patent for preparing *Limulus* amoebocyte lysate. He also started bleeding horseshoe crabs in his garage and sending free samples of the lysate to prospective clients. In 1977 the Food and Drug Administration issued him its first license to produce *Limulus* lysate. Associates of Cape Cod had completed its metamorphosis from a mom-and-pop real estate company to the world's first commercial manufacturer of *Limulus* amoebocyte lysate.

When *Limulus* lysate was first discovered, it was touted as a powerful new diagnostic test for gram-negative diseases. In 1976 an alarming medical crisis pointed to an alternate use.

A minor scientific report had come out that suggested that 1976 would be a bad flu year. Flu epidemics erupt from a pool of animal diseases in the Far East. Often the microbes mutate and start to infect humans. The "new" flu then races around the world, striking down millions of sufferers.

In 1976 it was suspected that swine flu would be the next candidate. Another candidate, President Gerald Ford, was informed of the report. He was advised of an opportunity to ensure his election: he could initiate the world's most massive free vaccination program to occur just before the November election.

Manufacturers were wary of the program so the government assumed liability. Private physicians were leery of the program so it was decided that it would be provided at government clinics. A senior adviser in the Food and Drug Administration was hesitant so he was fired. The order went out that the vaccinations were to be completed by the election. Before the much ballyhooed program got very far hundreds of people were dead from the vaccine and Jimmy Carter had been elected president.

By the end of "the great swine flu fiasco," no human cases of the disease had been discovered. The vaccine was proved to be ineffectual in preventing flu in swine.

The public remains wary of worthwhile health programs and manufacturers remain wary of producing vaccines and insurance companies are wary of selling them manufacturer's insurance. An embarrassed federal government is still thrashing its way out of a morass of litigation that may eventually make the Agent Orange scandal pale in comparison.

If there was any hero to arise out of this political maneuvering and ill-advised science, it was the horseshoe crab. Batches of swine-flu vaccine were tested with both the rabbit serum test and the *Limulus* amoebocyte lysate test. In the end it was *Limulus* lysate, not live rabbits, which proved that some of the vaccine was contaminated with endotoxins.

Following the crisis the Food and Drug Administration drew up regulations requiring that all influenza vaccines be tested for endotoxins prior to release. Shortly after, the FDA also published guidelines for the use of *Limulus* amoebocyte lysate in testing drugs, blood products, intravenous fluids, and disposable pharmaceutical devices like syringes. In 1983 the *Limulus* amoebocyte lysate test was accepted in the U.S. Pharmacopeia, the federal government's official registry of drugs, and pharmaceutical companies started dismantling their live rabbit colonies at a savings of millions of dollars.

The lysate test is produced by bleeding large female horseshoe crabs. The blue blood, caused by hemocyanin cells, is then centrifuged to separate the amoebocyte cells from the blood plasma. Distilled water is added to the cells, causing them to explode from internal pressure. Finally the lysate is freeze-dried into a fine white powder.

The test is simplicity itself. A small amount of *Limulus* amoebocyte lysate is mixed with an equal amount of the fluid to be tested. After an hour of incubation, the lysate will clot if endotoxins are present in the sample. The test is rapid, inexpensive, and easy to perform.

Although touted in the early scientific literature as a potential diagnostic tool for bacterial disease, the test has not panned out as well as had been expected. One reason is that the test cannot differentiate between various species of bacteria. Thus the doctor does not know which antibodies to prescribe.

There has been one celebrated case in which *Limulus* lysate was used experimentally to diagnose gram-negative spinal meningitis. Normally such a bacteriological diagnosis would take up to forty-eight hours to perform, but with the lysate test it requires only fifteen minutes, a life-saving margin with this fast-acting, lethal disease. The test has also been used experimentally in the diagnosis of gonorrhea, infections of the eyes and joints, and for endotoxemia in newborns.

Few funding agencies could have prophesied this outcome when researchers first started studying horseshoe crabs over a hundred years ago. Today the *Limulus* lysate story stands out as an example of the unexpected and beneficial results that can occur from funding basic research. Not only has the early research led to a new pharmaceutical industry but it is also creating a new marine industry—the aquaculture of marine research animals.

Yet it was only in the 1970s that Senator William F. Proxmire, flamboyant watchdog of government spending, awarded one of his golden fleece awards to Carl

Schuster, one of the pioneer researchers on horseshoe crabs. Perhaps someone should tell the good Senator the story of the horseshoe crab and present him with a leaden *Limulus* "for short-sightedness in the pursuit of governmental economy."

PART THREE

AUTUMN

Marine Animals
and the Human Brain

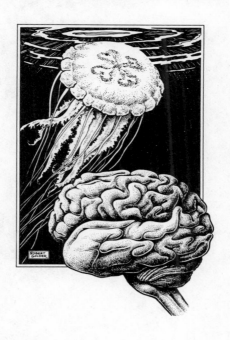

12

The Migration

It is night. The first full moon of autumn flickers through the greenish waters. The female horseshoe crab must hasten on her long journey offshore. She has reached the inlet where the cool saline waters of the Atlantic meet the warmer waters of the estuary.

Now she must make her way through the current-scoured runnels that lie between the beach and the sand-bars beyond. Here she joins a vast assemblage of under-water creatures migrating down the East Coast.

The water is replete with gelatinous bodies. Some belong to the ctenophore, *Mnemiopsis leidyi,* and some to the giant lion's mane jellyfish.

Mnemiopsis, a comb jelly, has eight rows of cilia combs running the length of its body. By day the cilia cast iridescent rainbows of color; by night they emit green flashes of bioluminescence. Flowing majestically through the greenish water, they sweep it clean of plankton.

The lion's mane jellyfish, *Cyanea capillata,* the largest jellyfish in the world, can reach up to eight feet across and trail two hundred feet of venomous tentacles. Tiny sparks of bioluminescence flash as microscopic animals blunder into its stinging cells (nematocysts). Larval haddock and butterfish are more fortunate. They seek out the tentacles for protection, darting in and out of this deadly tangle without discharging the cells.

As it pulses slowly through the water, the jellyfish resembles an eerie underwater space colony. Rather than a single animal, this pulsating body is a colony of polyp- and medusa-like individuals called "persons." Each "person" is considered a separate animal specialized for feeding stinging, reproduction, or propulsion.

Their means of reproduction is equally complex, though primitive. The bell-shaped medusa is the sexual adult form of the colony. Some colonies are male, some female. They will release sperm and eggs into the water and tiny mobile planula will develop. They will drift about for a few days until they pick up chemical scents from the substrate. These scents signal a favorable spot. The planula descend and metamorphose into an inconspicuous polyp. Growing on rocks, seaweed, or the edge of a creek, they resemble half-inch flowers.

In the spring, the polyps undergo asexual reproduction. Their bodies lengthen and segment. Each segment is a tiny medusa bud stacked like saucers along its body stalk. Each saucer pulses with its own independent life

until one by one they break off and swim away as minia-
ture adult medusae.

Theirs is a short migration, a simple dispersion of young.
The planula drift where the currents take them, and
chemoreceptors indicate where and when the migration
should end. Their behavior has required little in the way
of evolutionary change, a simple form of locomotion and
a simple sense of taste or chemoreception.

Other, more recent animals have evolved more complex mechanisms to aid their migrations. These mechanisms are integrated by increasingly complex central
nervous systems. By investigating the senses that make
migration possible, scientists can pry open a window to
better understand the workings of the brain.

Swimming below the jellyfish is an ancient animal that
has evolved more complex senses to aid migration. A sin-
ister black tail looms out of the greenish waters, hanging
ominously from the massive body of a stingray, now lost
in the turbidity of the swiftly flowing water. It stops to
root in the sandy substrate, its powerful wings—five feet
from wingtip to wingtip—flapping toward the bottom.
Great clouds of sand mushroom off the ocean floor as it
searches for food.

The sounds of cracking and crunching reverberate
through the water as the ray's powerful jaws crush the
thick shell of a clam. Its huge body rises from the sand
and gracefully swims south, leaving an empty crater of
sand behind.

Dasyatis centroura, the whiptail stingray, is from an
ancient order of fish related to sharks. Like the jellyfish,
it must migrate, but it has developed more complex sen-

sory and locomotor skills than the jellyfish. Its eyesight, chemoreceptors, and sensitivity to salinity and temperature, coupled with powerful muscles, are integrated by a complex brain to help the ray's ability to migrate.

Not far from the ray's feeding site, subtle paths through the eelgrass indicate that a more recent fish has started one of nature's most spectacular migrations—a pilgrimage to certain death in the cold murky depths of the Atlantic.

Large black-and-silver American eels, *Anguilla rostrata,* have taken on their spawning color. Their sides are slick and silver, and their eyes have started to enlarge. They have stopped eating and will not do so again. Their summer buildup of fat will give them energy to make the long migration to the Sargasso Sea, south of Bermuda.

In the depths of the Sargasso Sea, millions of adult eels will congregate, mate, and die. Their offspring, tiny larvae called leptocephali, will passively drift in the grip of the Gulf Stream, which will carry them to the coast from which their parents came.

Because the eels must return to their parent stream, they have had to evolve some complex physiological adaptations. The leptocephali of the American eels drift passively in the Gulf Stream for just long enough to reach the American coast.

Many of their brethren, the offspring of European eels, will not have changed, for theirs must be a longer journey. Entrusted to the constancy of the ocean currents, they must remain a part of the slowly drifting plankton community for as long as three years. On reaching the streams of their native coast they will use their chemoreception to guide them upstream.

The eels developed this migratory behavior in response to continental drift. As the plates that carry Europe and America drifted apart, eels had to evolve mechanisms to guide them back to their spawning area in the Sargasso Sea. But there are other animals in these waters—striped bass and terns—that have developed migratory behaviors in response to a more recent phenomenon: the Ice Age.

The outgoing tide has scoured sand, algae, and bits of detritus off the ocean floor. These are carried in suspension with the longshore current. As the horseshoe crab crawls across the substrate, vibrations from her feet spread through the sand. Tiny iridescent heads pop out of the bottom. These are the bright eyes of sand eels, *Ammodytes americanus*. They hesitate for the merest instant before erupting from the bottom to join a closely knit school of their kin. Thousands of their greenish silver bodies swim synchronously through the waters, moving as though cells of a single organism.

Longshore currents sweep the school over a submerged sandbar into the trough behind. Suddenly the calm is broken. A school of striped bass waiting in ambush hurtle toward the hapless eels. The water erupts into swirls and boils. Within seconds a small flock of terns gathers over the churning waters. Their raucous cries and jostling bodies spread excitement along the beach. Other terns drawn by the calls flock in from all directions. There is a frenzy of anticipation as hundreds of these feisty but graceful birds hover and plunge into the green waters.

Now the sand eels are caught in a vicious crossfire of bass hurtling out of the depths and terns diving from the sky. Massacre is everywhere. The waters churn with dismembered bodies, blood, and oil. One by one the terns emerge from the frenzy, each with a seven-inch sand eel dangling crosswise from its bill. Thus laden, they fly unerringly across the sandspit toward their nests.

Scores of fledgling tern chicks line the shore impatiently awaiting their next meal. As soon as they are old enough, they come to the beach, shortening the distance their parents must fly. Whenever an adult approaches, all the chicks start to cry, stretch out their necks, beg, and display.

The adults fly in, looking from side to side, calling to locate their own chicks. Sometimes they are unable to find each other and must reconnoiter at their nest site seventy feet into the thick beach grass of the colony. When the adult locates its chick, it feeds it quickly, sometimes hovering only long enough to deliver its food before returning to the fishing ground. Hundreds of birds still wheel and turn overhead. The noise is deafening, the smell pungent. Many hover and dive at any intruder, often drawing blood in the process.

The nesting season is over but a few pairs of incubating terns remain. Their chicks will hatch too late to grow into adults before the migration and these chicks will starve when the colony leaves for winter.

Terns and striped bass represent two species, one aerial and one aquatic, which have evolved highly complex physiological mechanisms to guide their migrations. Let's look at each.

The nesting colony is made up of common, roseate, and arctic terns. Soon the arctic terns will embark on the animal kingdom's longest migration. Flying twenty-two thousand miles, the equivalent of traveling across the continental United States seven times, the terns will end up in Antarctica. But their flight will be indirect. In the fall they fly northeast to Great Britain to join the flocks of other arctic terns sweeping south from their nests on the arctic tundra. Together the flocks fly down the length of the European and African continents to their wintering grounds on Antarctica.

Striped bass were originally related to freshwater species. During the retreat of the glaciers, many of their ancestors were swept out of swollen streams into coastal waters. Natural selection favored those that could make the transition and return to fresher water estuaries to lay their eggs. Thus they were able to enjoy the benefits of both environments: the extensive food resources of coastal waters and the safety of the estuaries for their offspring. The ability to inhabit two environments worked admirably for thousands of years; now, however, striped bass are threatened because acid rain in their brackish water spawning grounds is killing their eggs and larval offspring.

Scientists have long pondered the evolution of migratory behavior. Why do animals make such arduous journeys and how do they do it? The answers are nearly as many as there are species, and scientists who study them. However, there is a comparable evolution of physiology and behavior which runs through their descent.

If the anadromous migratory behavior of striped bass allows them to lay their eggs in the relatively predator-free environment of estuaries, are there comparable advantages for the long-distance flying terns?

It is theorized that birds, like striped bass and other anadromous fish, started to evolve their present migratory behavior with the retreat of the glaciers. Gradually they migrated farther and farther north in summer to lay their eggs. Why did they do this? Why didn't they simply stay in the mid-latitudes? It takes a great deal of evolutionary change to invest in the physiological mechanisms that make migration possible. What are the advantages that make this investment worthwhile?

By flying from one pole of the earth to the other, terns expend a great deal of energy and take many risks. Why do they do this? There seem to be two answers: food and time. The sand eels of the north and krill-eating fish of the south provide them with a food source far more concentrated than they could find if they stayed in the mid-latitudes.

The second answer, time, is more complicated. Birds have had to make many compromises for their ability to fly. One of these is the vulnerability of their chicks. The period of highest mortality for birds occurs from the time the eggs are laid until the chicks can fly. Anything that birds can do to reduce this period will dramatically increase their chick's ability to survive and reproduce. By flying to the north in summer, nesting birds enjoy more hours of sunlight to feed their chicks. This decreases the amount of time their chicks are vulnerable. Terns, by flying to Antarctica, have the added advantage of the longer hours of the antarctic summer.

But how do animals navigate on these long migrations? People have developed many navigational tools, such as sextants, and chronometers. Sextants work by measuring the height of the sun, moon, and stars. Chronometers are simply clocks that tell mariners how far they are from Greenwich, England. By correlating the two—the height of a celestial body and the time of day—a mariner can pinpoint his location.

Are there analogies to sextants and chronometers in the animal kingdom? National Marine Fisheries scientists have noted an interesting cycle in striped bass. In the north, catches of striped bass seem to fluctuate over an 18.1-year period, which happens to coincide with the changing wobble or azimuth of the moon. If striped bass use the moon to cue their migrations, they would stay in the north longer when the moon approaches its highest point. This would give the fishermen a longer time to catch bass and could account for the higher catches during these years.

Horseshoe crabs also use their simple eyes to receive ultraviolet light from the moon. Could it be that the female horseshoe crab is also using the moon to navigate and as a cue to trigger her migration? If this is so, then all she would need is a timepiece to navigate accurately. Such a timepiece is the biological clock centered in her brain. Could it be that light-sensitive organs and internal biological clocks evolved in tandem like sextants and chronometers to aid in animal migration?

In the following chapters we will look more closely at the neurological mechanisms that make complex behaviors like migration and mating possible.

13

A Short Course
in Neurobiology

It is early autumn. The shallow waters of the bay still retain the heat of summer but the days are perceptibly shorter. The female horseshoe crab has returned from her unanticipated odyssey through the biomedical industry. She must now gorge to replenish her blood and prepare for the autumnal migration back to the deeper waters of the continental shelf.

She crawls slowly through swaying fronds of eelgrass beds. Thousands of underwater acres of eelgrass glisten with silvery oxygen bubbles. The sun is providing the energy for their photosynthesis. It is this constant influx of oxygen that helps to keep these shallow waters producing food for animals like the horseshoe crab.

Chemical receptors along the crab's carapace pick up the faint traces of potential food. Ahead of her the uplifted head of a feather-duster worm, *Myxicola infundibulum,* is raised a few centimeters off the muddy substrate.

Suddenly the worm senses the crab's presence. Within a millisecond it disappears. Only a nondescript blob of jelly remains on the muddy bottom. A rapid electrical impulse has fired down the length of the worm's body, triggering the response. The impulse traveled down a giant axon, the long process that runs from the cell body to the nerve ending.

Neurobiologists have studied marine worms to understand precisely how rapid electrical impulses are generated and pass down the nerve. They have found that the nerve axon contains channel molecules that act like gates permitting charged ions to flow into or out of the cell membrane. The difference in positive and negative charges across the membrane generates an electrical impulse and passes it down the length of the axon, triggering the muscles to retract.

The use of electrical impulses is the most direct and rapid means of getting information from a central processing area like the brain to the peripheral muscles. It is a system that evolved early in the primitive marine invertebrates, but it is as complex as our most advanced computers, and it is only one method that nerve cells have evolved to transmit information to and from the environment.

Despite the efficiency of the worm's giant axon it proves to no avail. The horseshoe crab quickly dispatches the tasty morsel and moves on to further food, this time devouring the fragile-looking *Gemma gemma* clam, a miniscule mollusk less than a quarter of an inch across. The tiny white and purple clam looks exactly like an immature quahog, *Mercenaria mercenaria,* a similarity that has led to an unfair discrimination against horseshoe crabs. Shellfishermen have understandably, though falsely, accused the horseshoe crab of feeding on quahogs, the mainstay of true New England clam chowder.

Horseshoe crabs do feed on immature soft-shelled clams and razor clams. It is unknown how they can feed on razor clams before the clam digs below the substrate, although it's possible that the crab releases a toxin that paralyzes the clam's large fleshy foot, preventing its escape. After the horseshoe crab masticates the mollusk with hundreds of tiny bristles on the base of her legs, all that remains is a small pile of broken shells littering the ocean bottom.

Although encounters with mollusks have given the horseshoe crab an unjust reputation within the shellfish industry, there is a more deadly organism in these waters. It is the minute dinoflagellate, *Gonyaulax tamarensis,* whose neurotoxins produce paralytic shellfish poisoning or, to use the more colorful term, red tide.

Red tide is a new menace in Cape Cod waters. It spread south from Canada in 1972 when Hurricane Carrie poured so much rainwater into New England coastal waters that salinity dropped enough for massive blooms of *Gonyaulax* to spread south. Now they have a foothold in brackish waters from Maine to Cape Cod. Recently they have even spread around the sandy Cape Cod barrier to infect shellfish as far south as Long Island Sound.

The red-tide toxin is a powerful neurotoxin fifty times stronger than curare poison. The CIA and the KGB have both experimented with its use in James Bond-type assassination attempts. Not many years ago a Bulgarian defector was gashed in the leg by an umbrella-wielding Londoner. A few hours later the Bulgarian died from what appeared to be a massive heart attack. However, the autopsy revealed that embedded in the man's leg was a hollow steel ball that still contained the residue of red-tide neurotoxin.

Neurobiologists have used this toxin for more interesting, if less exotic, purposes. They have discovered that the toxin interferes with the calcium channels in nerve axons. These molecular channels create electrical impulses by pumping calcium ions into axon membranes. The use of neurotoxins in biomedical research has been a key to understanding the molecular basis for nerve mechanisms.

Hovering in the water just above the horseshoe crab is the animal primarily responsible for our present understanding of nerves and axons. It is the squid, *Loligo pealei,* one of the most advanced members of the mollusk phyla. The horseshoe crab moves closer, frightening the nervous yet curious squid. With an instantaneous dart the squid shoots back about three feet. The giant axon that activates this response has given the squid a worldwide reputation within neurological circles.

Recently biomedical researchers have started to use squid, horseshoe crabs, marine worms and red-tide organisms as models to understand the workings of the human brain, and prevents such illnesses as Alzheimer's disease. We will return to that work in Chapter 15.

14

The Age of *Aequorea:*
The Jellyfish that
Lights up the Human Brain

The silent waters of the autumn ocean shimmer in the light of distant galaxies. A small sailboat drifts aimlessly, and golden-green flashes of luminescence explode beneath it as small fish blunder into the gelatinous bodies of *Aequorea* jellyfish. Millions of their seven-inch umbrella-shaped bells pulse rhythmically through the water feeding on colonies of salp.

Converging surface currents have concentrated the plankton in this area. The salp, transparent tubes of protoplasm, have responded by reproducing wildly. Immense colonies of the simple tunicates stretch out for thirty feet. Each salp in the chain is a daughter clone of the one before. Such dense colonies of this simple animal have been known to stall ships by clogging their intake valves.

The jellyfish have responded to this unexpected patch

of food with their own reproductive orgy. A small cleavage is forming in the umbrella-shaped bell of one of the jellyfish. The two halves look like Siamese twins pulsing slowly in synchrony. Eventually two new medusae will break apart and swim off in opposite directions.

Below the surface a tentacle of one of the jellyfish has brushed against a salp. Immediately hundreds of the jellyfish's stinging cells explode and drive their sharp hollow spines into the soft flesh of the urochord. Like hypodermic needles, the stinging organs or nematocysts inject a drop of paralyzing toxin into the salp's body.

Chemicals from the wound pour into the water, triggering receptor cells in the jellyfish. Its tentacles commence to writhe and twist, slowly drawing the salp up toward its mouth. The mouth opens and soft lips slide along the salp's body, slowly engulfing it.

This deadly ballet between gelatinous animals has persisted for eons, for it was primitive animals like jellyfish that first evolved the crucial ability to detect and capture food. Simple automatic responses to specific chemicals caused these early predators to switch on the correct feeding behaviors.

It is sobering to think that the complex receptor sites on the nerves of the human brain are simply modifications of these chemoreceptor systems developed in coelenterates over seven hundred million years ago. It is these receptor sites which make possible learning, memory, and all the other magnificent emotional and cognitive abilities of the human brain.

Below the salps another jellyfish reveals how a primitive sensory system can be coupled with a simple network of nerves to coordinate a complex escape maneuver.

A current of water, perhaps from a passing fish, has tumbled one of the jellyfish. Now it pulses upside down, diving toward certain starvation in the ocean's cold, dark planktonless depths.

But all is not lost. Pulse by rhythmic pulse the overturned bell of the jellyfish slowly rights itself. Now each contraction returns it to the plankton-rich waters of the surface.

Statocysts, small organs scattered along the rim of the jellyfish bell, have helped avoid disaster. Each hair-filled chamber contains a tiny granule of calcium carbonate. When the jellyfish was overturned, the granule fell against some of the hair cells, sending an electrical impulse to the animal's nervous system, a simple array of nerves that coordinates the jellyfish's slow rhythmic contractions.

Statocysts were originally thought to be organs for hearing. In 1893, however, the Austrian scientist Kriedl conducted an elegantly simple experiment to reveal their true function as organs of equilibrium and orientation. He was studying some shrimp that use sand grains to trigger their statocysts. Each time the shrimp molts, it must replace the old sand grains lost in the shedding process.

Kriedl removed all the sand grains in the shrimps' aquarium and replaced them with iron filings. After the shrimp had taken in the filings, Kriedl could make the hapless animals assume any posture by simply moving a magnet outside their aquarium. The shrimp had lost their ability to tell which way is up.

Someday the ability to tell which way is up may win a jellyfish entry into one of the world's most exclusive clubs—that select group of men and women who have gone into space. The common moon jellyfish, *Aurelia aurita,* has already taken a simulated space flight by riding in a test tube tucked in an astronaut's shirt pocket. The celebrated jellyfish who does ride aboard a spaceship may help scientists learn why astronauts get space sickness and lose their ability to keep their balance for a few days after splashdown.

On earth, jellyfish are helping mankind explore an equally intriguing frontier, that of the human brain. We are being aided in this quest by one of the jellyfish's most fascinating attributes: bioluminescence.

Bioluminescence has long been a curiosity of nature. During World War II pilots used the glowing wakes of ships to track submerged submarines and to guide them back to their own aircraft carriers at the end of their missions.

As far back as the first century A.D., Pliny the Elder described the biological nature of bioluminescence. He reported that the luminous slime of one of the Mediterranean jellyfish was thought to have medicinal value when consumed with wine or rubbed on someone with fever. While the medicinal value was suspect, one can imagine that the heavy pyrotechnics involved in drinking glowing wine and having one's body shine like a god's was enough to raise the spirits.

Apparently Pliny's early understanding of the biological nature of bioluminescence was lost for seventeen hundred years. Descartes believed that when waves hit a ship, salt particles separating from the water generated sparks similar to those emitted by flint. It was only during the great exploratory cruises of the late nineteenth century that the biological nature of bioluminescence was verified. Since then scientists have discovered that an impressive array of animals, including bacteria, dinofla-gellates, jellyfish, ctenophores, squid, insects, and fish, have independently evolved bioluminescence.

Several species of animals use bioluminescence to attract food. Deep sea angler fish have evolved modified dorsal fins that dangle in front of their mouths. On the end of this fin is a luminous lure resembling a bioluminescent copepod. When small fish approach to investigate the lure, they fall prey to the angler fish.

In New Zealand, glowworms dangle sticky threads beaded with bioluminescent globs from the ceilings of caves. Flies attracted to the light stick to the threads and are drawn in and devoured by the voracious glowworms.

Other animals use bioluminescence for defense. The Indo-Pacific flashlight fish has twin compartments below each eye. The compartments are filled with luminous bacteria that can be extinguished by closing a lid. To escape an enemy, the small fish simply flashes one eye compart-ment and then swims off in the opposite direction. This wink-and-run tactic has made flashlight fish a favorite attraction in major aquariums around the world. Deep-sea squid employ an equally ingenious defense. They squirt a cloud of luminescence into the darkness and jet propel themselves to safety.

Finally many animals use bioluminescence for sexual attraction. Throughout the world fireflies emit distinct patterns of flashes to attract mates of their own species. In Melanesia they have developed this behavior even fur-ther. Thousands of male fireflies gather in specific trees and flash in unison like a giant living Christmas tree. The spectacular display attracts females from miles around.

Only recently has mankind unraveled the biochemical reactions underlying bioluminescence and started to harness it for medical purposes. In hospitals anesthesiology is still considered an art, rather than a science, because doctors must constantly monitor and alter the dosage and mix of drugs to keep their patients on the fine line between unconsciousness and death. Most deaths that occur during surgery are caused by incorrect anesthesiology rather than by the operation itself.

But in the future, luminous bacteria that dim when they come in contact with drugs may give anesthesiologists a rapid and sensitive means of evaluating the potency and dosage of anesthetics during an operation.

Extracts from fireflies are presently being used to test for adenosine triphosphate (ATP), the energy molecule that underlies human metabolism. Commercial kits of the bioluminescent system are used clinically in many laboratories to test for heart disease.

But it is the *Aequorea* jellyfish that has given mankind an exquisitely sensitive probe to explore the workings of the human brain. A single molecule of aequorin, a protein extracted from the *Aequorea* jellyfish, emits one photon of light when it is activated by calcium. This is crucially important because the human body uses small amounts of calcium to activate muscles, glands, and nerves.

By injecting aequorin into a nerve, researchers can "see" calcium as it flows into nerve membranes, triggers electrical impulses, and causes neurotransmitters to be released across a synapse from one nerve to another. All this can be done without disturbing the cells involved.

Thus the *Aequorea* jellyfish, one of the most primitive

animals, has given us an unprecedented opportunity to visualize the higher processes involved in memory and learning. For the first time mankind is at the threshold of watching himself think. Pliny the Elder would be pleased.

15

The Brain, the Nerve, and the Squid

There is no more wondrous creation in the known universe than the human brain. Bathed in the slightly saline solution which is the hallmark of its early evolution among marine animals of the Cambrian, the human brain represents the highest achievement of our planet's evolution. This billion year old process has created, altered and improved an organ of immense complexity, discrimination, subtlety, and flexibility.

At the core of our wondrous computer are the neurons, the microchips of the human brain. Neurons appeared in evolution six hundred million years ago among coelenterates, primitive jellyfish that swam rhythmically through the shallow waters of the Cambrian seas. It is for this reason that neurobiologists return to primitive marine creatures to understand the workings of neurons, the cells that generate the emotions, ideas, and aspirations upon which our humanity is built.

Starting with a single cell in the developing embryo, the human brain grows quickly. Tiny protein factories in our cells produce endless complex chains of sensory and motor cells: the advanced network of neurons, dendrites, axons, and synapses that make up the interconnected circuitry of our central nervous system.

It is this intricate biological computer that filters and translates the sights, sounds, and impressions of our environment into the emotions and ideas that dictate human behavior. There is no more pressing endeavor than to unravel the workings of our brain before one of our kind, misguided by our primate heritage, pushes the button that will end the further evolution of mankind, the human brain, and culture.

Until the past few decades mankind has had little understanding of the inner workings of the brain. Philosophers have debated the differences between the brain and the mind. Pioneering thinkers like Freud have probed the outer manifestations of normal and abnormal thought processes and pharmacologists have stumbled upon drugs that enhance, mimic, or alter our consciousness. But it is only during the past thirty years that scientists have been able to unravel the workings of the brain on a cellular level; that knowledge is having repercussions in medicine, psychology, and philosophy.

The modern quest to understand the human brain has replaced gene splicing as the most competitive, challenging field in biology. Among its pioneers are the greats—Pavlov, Freud, and Skinner—plus a host of lesser known scientists. Almost forgotten in this pantheon of giants are some of the animals that have brought us this far—the millions of lobsters, squid, monkeys, snails, and horseshoe crabs that have been sacrificed for our knowledge. In fact, one animal, the common squid, *Loligo pealei,* has had such a long-standing and significant history in neurobiology that it has been suggested that it deserves a Nobel Prize.

My first view of the *Loligo* squid gave me a vivid introduction to the reason for its usefulness. I was diving with Ken Read, a close friend and mentor from Boston University's biology department.

We were diving on the incoming tide of Cape Cod's Pleasant Bay. The current held us in its gentle grip, gliding us effortlessly through the sun-dappled waters. Sea robins, lady crabs, and an occasional dogfish skittered over the sandy bottom. Suddenly we looked up and saw a lone foot-long *Loligo* squid suspended in the water. It stared at us with large intelligent eyes while waves of nervous color undulated down the length of its body. The colors were controlled by a network of nerves innervated into the squid's skin muscles. Tiny pulses of electricity coursing through these nerves caused the muscles to open and close color filled chromatophores embedded in the squid's skin.

We slowly lifted the camera to record this multihued phenomenon. With split-second brilliance the strobe light flashed and the squid disappeared. All that remained was empty green water.

Glancing around, we saw, four feet behind its former location, the same squid blanching solid white with apprehension. Slowly, row upon row of pinks, browns, and reds again undulated down its body and the squid allowed us to swim in closer for another picture. Again and again we tried the simple experiment. Each time the squid shot back faster than our eyes could register.

The squid had jet-propelled itself backward by contracting its mantle, forcing water to squirt out of a siphon below its body. The nerve that triggers this instantaneous reflex has a giant axon, a long fibrous process running the length of the squid's body, connecting the animal's primitive brain to its muscles. The giant axon is similar to the human sciatic nerve, with its long axon traveling from the human brain down the spinal column to the toes.

Despite its shorter length, the squid's giant axon is many times thicker than a human's. It was the discovery of this giant axon, a large white fiber often as thick as a pencil lead, that helped create the field of neurobiology.

Until the summer of 1936 most scientists thought the squid's giant axon was a blood vessel. An MBL scientist, Dr. J. Z. Young, after reading a manuscript which postulated that the process was in fact an unbelievably large axon, decided to test the hypothesis. He contacted Dr. H. Keffer Hartline, a colleague who was perfecting electronic equipment for studying the optic nerve of horseshoe crabs.

After the lab was closed, the two scientists decapitated a large *Loligo* squid, slit open its mantle, and teased apart the five-inch fiber. They put the fiber in a saline water bath and attached an electrode to either end. A jolt of electricity registered on their oscilloscope. The spike was so strong that Dr. Hartline, fearing for his equipment, raced down the hall thinking that it had been caused by the nearby X-ray lab. But the lab was closed and locked for the night. The jolt had come from the squid's giant axon.

The discovery of this fine plump axon led to a cascade of new information about the electrical and biochemical nature of nerves and synapses, the crucially important gaps between neurons.

For many summers following Young's findings, scores of researchers descended on the Woods Hole MBL to push, prod, and inject thousands of squid axons with electrodes, voltage clamps, poisons, drugs, and radioactive tracers. Throughout the years researchers have discovered that electrical impulses are generated by the negative and positive charges of ions inside and outside nerve membranes. They have shown that these ions of sodium, potassium, and calcium can be pumped through channel cells to alter the electrical potential of the nerve. Often these changes seem to correspond to different types of learning.

For instance the squid that Ken Read and I photographed gradually became accustomed to our strobe light, a type of learning scientists term habituation. They believe that during habituation calcium activates the enzymes that control the channels through which the ions flow, inhibiting the cell and reducing the electrical charge through the nerve. The squid responds less and less to sucessive flashes. It has "learned" that a strobe light will startle but not harm it, and the altered cellular pathways in the nerves are the mechanism for that short-term learning.

If we had continued flashing the strobe light at the squid for several days, structural changes would have occurred in the synapses between neurons. More areas would have developed to release neurotransmitters, which control the amount of time these cellular channels remain open, and this in turn would have altered the electrical discharge of the next nerve. Long-term memory would result.

In much the same way, scientists believe that learning a new motor skill such as riding a bike or playing tennis requires practice until changes in the cellular channels and synapses occur. These changes alter electronic and chemical pathways so that we retain such motor skills for life.

The similarity of squid axons to human axons extends to tiny organelles and axoplasm, a viscous substance through which nutrients and neurotransmitters flow. Scientists have only recently discovered new groups of neurotransmitters. Because the neurotransmitters carry messages to the synapses at the end of the nerves, where they are released to carry messages to adjoining nerves, their transport has proved to be crucially important to human health.

Two prominent diseases caused by the impairment of axonal transport are Alzheimer's disease and amyotrophic lateral sclerosis (or Lou Gehrig's disease). Alzheimer's disease is a devastating illness that causes premature progressive senility. Victims of the disease lose the ability to speak, remember, eat, drink, and remain continent. While passing through these devastating changes they remain conscious and are frustrated by their progressive deterioration. When death finally occurs, it is often a guilt-ridden relief to the entire family which has had to cope with caring for the victim.

Scientists are convinced that in Alzheimer's disease

irregular tangles of neurofilaments and twisted microtubules interfere with the passage of signals from cell to
cell, resulting in progressive memory loss. Lou Gehrig's
disease involves the deterioration of motor cells, leading
to progressive loss of body control. Several researchers
are presently studying squid to identify a specific molecular therapy, perhaps a drug, that will improve defective
axonal transport in victims of these two diseases.

There is one cloud on this otherwise rosy picture. The
MBL in Woods Hole is the only laboratory where sufficient quantities of live *Loligo* squid are available to do
extensive axon research. But Woods Hole scientists are
not the only people interested in squid. A new, expanding squid fishery has developed along the southern Massachusetts coast. Every spring during April and May scores
of boats from distant harbors descend on southern Cape
Cod ports to fish for spawning squid.

Recent catches have been disappointing, and many
people point to overharvesting of spawning squid as the
reason for the poor landings. Neurobiologists fear that
vital nerve research could be jeopardized by the expanding commercial catch. The MBL has even suggested that
a squid sanctuary be established in Nantucket Sound. The
proposal is sure to raise the ire of commercial fishermen
who have come to rely on the spring squid harvest as a
way to stretch out their fishing season, especially when
the weather is unpredictable offshore on Georges Bank.

While sanctuaries and regulations may provide short-
term solutions for ensuring steady supplies of marine
research animals, the long-term but expensive solution
may be the development of mariculture facilities to raise

animals like squid and horseshoe crabs. This solution will have the added benefit of providing research animals with known genetic lines.

Whatever the outcome of this present conflict, the "Woods Hole squid" has put in many years of service in helping unravel the intricacies of the human brain. Steps should be taken to see that it gets the recognition and protection it deserves.

16

The Eyes Have It

As a child, one of my passions was watching horseshoe crabs. I would observe them during the day and dive with them at night. When spending so much time with the creatures, you notice that during the day they appear totally blind. At night however, they seem far more perceptive.

When the moon is full and the female is preparing to lay her eggs, the males flock in from all directions. Biologists have long debated whether vision or pheromones are responsible for this mating behavior.

Formerly I was convinced that this was caused by pheromones, chemical aphrodisiacs that are a common form of communication among invertebrates. Lobsters use them in mating and the humble female Cecropia moth employs them to attract males from as far away as three miles. They are even believed to be the basis behind the "chemical attraction" between humans. The perfume industry has spent millions of dollars trying to isolate an actual human pheromone.

It is one of the ironies of my early life that had I only paid more attention to my sister's love life I would have learned a great deal more about horseshoe crabs and would not have supported the wrong side in the great vision-versus-pheromone argument.

It was only recently that I discovered my loss. I was telling my long-suffering parents about the horseshoe crab research of the late Dr. H. Keffer Hartline when my mother piped up, "Oh, wasn't there a Dr. Hartline who spent the summer in the Benton's cottage? I think he was studying something about horseshoe crabs. Remember Fay [my older sister famous for her frequent and intense love affairs] had a crush on Danny Hartline."

"Good God, you mean that H. Keffer Hartline was living next door and I wasn't even aware of the fact!" I was growing indignant about my sister's sense of priorities.

The memories of that fateful summer came flooding in. All we heard about for the better part of the summer was this dashing Danny Hartline, who would churn up my sister's already hyperactive hormones on long midnight sailing trysts. There she was, hanging around his house trailing a scent of undoubtedly powerful pheromones, totally oblivious to the pioneering horseshoe crab research going on right under her nose!

Of course it turned out that not only was Dr. Hartline doing "some kind of research on horseshoe crabs," but he was well on his way to earning the 1967 Nobel prize for his work on horseshoe crab vision, research that still elucidates much of what we know about the mechanisms of neurobiology and human vision. Since that single moment of epiphany I have made it a point to closely examine my sister's rich and varied love life. It still remains a fascinating subject, though so far, I have not been able to assay her pheromone output.

Dr. Hartline chose the horseshoe crab for several reasons. He was interested in the evolution and mechanisms of vision, so it made sense to select a primitive animal. More importantly, however, the horseshoe crab is an ideal research animal because it possesses one of the largest, most accessible optic nerves in the animal kingdom.

During the early stages of its development the horseshoe crab's compound eyes lie quite close to the brain. As the crab develops, the brain becomes what is called a circumesophageal ganglion, meaning the brain lies in a ring around the animal's mouth and throat. (In other words, the horseshoe crab eats through its brain.)

Simultaneously the eyes migrate to the sides of the carapace and two long optic nerves grow back to the brain. Not only is this nerve long, up to four inches in a large adult crab, but it lies just below the surface. One must only cut a small hole in the carapace to expose the nerve and then attach electrodes to the nerve to record the electrical impulses flowing back and forth to the brain. Once the electrodes are in place the researcher can experiment with the crab's visual system. As uncomfortable as it might sound for the hapless crab, researchers insist this procedure will not unduly affect it. Many crabs have lived for over a year after undergoing this procedure.

Working with an excised eye, Dr. Hartline first discovered that strong electrical impulses surge down the optic nerve. He also found that a complex system of circuits lies below the photoreceptor cells. These circuits enhance horseshoe crab vision by altering the signals sent to the brain. Thus, if one photoreceptor cell is receiving bright light and the cell beside it is receiving dim light, the first cell inhibits the signal from the second. The horseshoe crab does not receive a true image but an enhanced image with better contrast between borders.

Our brain does the same thing when we look at the horizon. The water at the edge of the horizon is no darker than anywhere else, but interaction between cells causes us to see the border more clearly by making the water at the horizon appear darker and the sky above it appear lighter. The process is called lateral inhibition and it enhances our ability to discriminate borders. This has a distinct survival value for us and for horseshoe crabs. (One curious application of this finding is that X-ray technicians are now taught about lateral inhibition so they will not misinterpret X-rays by seeing hairline fractures where none exist.)

Dr. Hartline always told his students to keep things simple, to "avoid vertebrates because they are too complicated, avoid color vision because it is much too complicated, and avoid the combination because it is impossible." His students have gone on to disobey most of his dictates thinking perhaps that their mentor had discovered everything there was to know about invertebrate vision.

But one of his students, Dr. Robert Barlow, stayed with horseshoe crabs and found that Dr. Hartline made one mistaken assumption: he thought horseshoe crab vision was simple. As Dr. Barlow has shown, horseshoe crab vision is uniquely complex.

One of the most noticeable traits of mating male horseshoe crabs is that they consider anything—old boots, bricks, anchors, boats, feet—anything in the water to be a mating female. I became acutely aware of this while helping to film horseshoe crabs for the NOVA film, "The Sea Behind the Dunes." Not only did the crabs clamber over our camera tripods but they attempted to mate with our feet. It is very difficult to hold a camera steady when thirty or forty lascivious male horseshoe crabs are tickling your toes.

Dr. Barlow proceeded to study this phenomenon more quantitatively. Night after night he placed concrete horseshoe crab models along the shore and kept track of whether the males preferred white models, black models, square, round, or cubic ones. Dr. Barlow discovered that the crabs discriminate between colors and shapes, and that they do this through the interactions of a biological clock located in the brain, their complex eyes, and their small simple eyes.

Signals from the simple eyes alter the structure of the compound eye so that lateral inihibition is muted. This diminishes the eye's ability to see borders but enhances its ability to see objects in dim light. The environmental cue that sets off this mechanism is ultraviolet light, which is most prevalent when the moon is full. When is the moon full? During the high-course spring tides, the exact time that the female must come ashore to lay her eggs. A most elaborate mechanism for such a "simple" creature!

If you dive at night you will see how effective a signal

the full moon can be: At a depth of twenty-five feet it still casts enough light to read by. The horseshoe crab's vision system developed long before the evolution of rods for night vision. Because the horseshoe crab possessed only a single type of photoreceptor cell, it evolved a complex interaction to ensure that it came ashore to mate exactly at the peak of the high-course spring tides. But was this interaction evolved to help the crab see or to help the crab know when to lay its eggs?

When the horseshoe crab migrates offshore in the winter, it digs down into the sediments. Here it is isolated from most environmental cues like changing light or temperature. In the laboratory horseshoe crabs have been kept in the dark for an entire year and their biological clocks have continued to function accurately, with the crabs maintaining their usual nocturnal rhythm of activity. Could it be that the simple eye's reaction to ultraviolet light from the full moon activates the crab's biological clock so that it knows when to start migrating shoreward at the end of its hibernation?

It seems that the horseshoe crab evolved this complex interaction between the brain and the eyes more to help it keep track of time than to help it see.

In the next chapter we will investigate how the interaction between the brain and the ear has affected human civilization. But do horseshoe crabs use vision or pheromones for mating? For now let's call the great debate a draw. As in human mating, they probably use both.

17

Hearing, Whales, and Civilization

Human civilization depends on 23,500 tiny hair cells lining the tunnel of Corti in the inner chamber of the human ear. Without these nerves there would be no hearing, no language, no music, no civilization. Hearing is one of the most complex of all senses and one that is surprisingly rare in the animal kingdom.

Hearing has a venerable heritage that goes back to a simple sensory system, the lateral line organ, visible as the dark line that runs down the length of a fish's body. Evolved during the early Devonian, the lateral line has gone through a series of adaptations leading to an impressive multitude of senses, the detection of pressure, weak electromagnetic stimulation, and hearing. To gain an understanding of these adaptations we must only look into the autumn waters of Cape Cod. . . .

Thousands of silversides, *Menida menida,* form a dense school. Multitudes of tiny hair cells within a labyrinth of pits along their lateral line system detect minor changes in water pressure. This helps them maintain an equal distance from each other in the school.

Suddenly a new vibration frightens the fish at the periphery of the school. A smooth dogfish, *Mustelus canis,* approaches. Hair cells in the lateral line of the silversides generate electrical pulses that travel down nerve fibers to a large complex cell, the Mauthner cell. The Mauthner cell is a specialized command cell that quickly relays further electrical signals to motor neurons connected to muscles in the silversides' tails. Each fish rapidly flips its tail to the side.

A starburst of fusiform bodies explodes in front of the shark. This rapid, automatic startle response transforms the orderly school into a chaotic explosion of silver missles, each hurtling in a confusing pattern of escape. The silversides have used their lateral line's ability to sense vibrations in water both to maintain their school and to escape the dogfish. The Mauthner cell transforms this information into a rapid automatic escape response, a kind of fixed-action pattern.

Many animals have evolved similar fixed-action patterns triggered by specific signals. Frogs have "bug detectors" that cause them to strike at any small object moving rapidly across their field of vision. Many fish strike at a flash of silver or a slash of red much quicker than they would at the colors of a natural minnow. This is why the best fishing lures have the most exaggerated colors or bizarre movements, not the ones that, to our eyes, look most like a live minnow.

Likewise, young herring gulls automatically peck at the orange spot on their parents' beaks to elicit their parents' feeding response. They will do the same thing to the most outrageous cardboard model of an adult seagull as long as it has an orange spot.

Some human males have even been accused of having automatic fixed-action responses to a provocatively lifted female skirt. Female pickpockets have long used this stratagem to distract their male prey while an accomplice artfully heists his wallet.

But this time the dogfish has little interest in such fast prey. It cruises slowly toward the sandy bottom, sinuously weaving from side to side. Suddenly it plunges headfirst into the sand and rises with a flat flounder wriggling crosswise in its snaggle-toothed jaws. It has used its ability to detect electricity to catch its meal. Ampullae of Lorenzini, a series of pits in the shark's head, and lateral line system were able to detect the faint electrical aura emitted by the flounder even as it hid beneath the sand. The sharks' electroreception abilities are among the most sensitive in the animal kingdom. They can detect an electrical current equivalent to what would flow between the poles of a flashlight battery even if the poles were 900 miles away from each other.

Unlike the terrestrial environment, the underwater world contains a symphony of electrical and electromagnetic waves. The sharks' ability to detect such faint emanations give them yet another weapon in their formidable arsenal of predatory advantages.

Sometimes such advantages backfire, such as when sharks attack propellers and the bottoms of metal boats. Just before they attack they roll back their eyes and substitute their vision and keen sense of smell with their ability to detect electricity. Often such reliance leads them to by-pass a soft juicy chunk of bait and slash into the metal parts of a fisherman's boat, which is giving off a far more tantalizing electrical signature.

Other fish have evolved specially adapted cells for generating electricity. I still remember the shocking experience of reaching into a thrashing pile of fish just unloaded onto the deck of a research vessel. By chance I grabbed the tail of a torpedo fish known for its ability to generate electricity. The jolt knocked me backwards even though most of the fish's electricity had been discharged in contacts with other fish in the crowded net.

Electric eels are the prime producers of electricity in the animal kingdom. They posses the ability to kill a man with a powerful jolt of electricity. Even the lowly catfish have cultivated the ability to communicate to each other about sex and territory through faint electrical signals.

My favorite electricity generating fish is the stargazer, which lies on the bottom with his eyes looking up and his mouth open. When its prey swims overhead, the star-gazer merely emits a jolt of electricity that stuns its prey, knocking him into the waiting jaws of the decidedly lazy stargazer.

On the surface, other animals use sound to gather their food. A large pod of pilot whales have driven a school of squid into the shallow waters on the inside of the arm of Cape Cod. The whales are using echolocation to find and kill their prey. Like a submarine using sonar, they are sending out pulses of sound that bounce off the squid and return to their sensitive internal ears. As the whales approach the squid, they might even emit a deafening

sonic blast to stun the mollusks for easier capture. Sometimes mother whales use their sonic abilities to detect gas
in their infants, using sound waves in much the same way
as an obstetrician scans a mother's womb with an opti-
cally enhanced ultrasound device to give a picture of the
developing fetus.

But now the tide is receding across the many miles of
Cape Cod Bay's sand flats. Echolocation, which works so
well at locating prey, is unable to discern the shallow
gradient of these shoal waters. The pod cannot tell which
direction will take them back to deep water.

The bay fills with frightened squeals and alarm calls as
mothers and infants sense their plight. Some human
helpers have arrived on the scene to try to guide the
frightened whales back to deeper water. With stubborn
resolve, however, the whales keep returning to the same
beach. Within a few hours it is over. Sixty-five sleek black
whales heave failing gasps of breath, the weight of their
bodies crushing their lungs and internal organs. Heat
trapped by their blubber is overheating their tissues. Large
sad eyes look one last time at stranded family members
as one by one the magnificent creatures die.

This scene has repeated itself on this same beach for
generations. Thoreau wrote of it in 1855 and it has been
reported almost yearly during the pilot whales' annual
autumn migrations. What is it about this beach that causes
these mass strandings?

Here is a fragile peninsula of sand that hooks out far
into the Atlantic Ocean. In geological terms this peninsula is one of the newest formations on earth, left here
after the retreat of the glaciers a scant twelve thousand
years ago. But whales have been making their offshore

migrations for millions of years. Could it be that whales guide their migrations by using some sense unfamiliar to humans, and that this sense evolved millenia before the formation of this recent caprice of geology we call Cape Cod?

Other animals use the earth's magnetic field to orient themselves during their migrations. Sharks and skates have been trained to respond to the earth's magnetic field. Pigeons are known to use magnetic detection when the weather is overcast or when they can't use the sun for orientation.

Scientists at Woods Hole discovered a species of bacteria that always swims north along the lines of the earth's magnetic field. Because they live in the Northern Hemisphere, this also ensures that they swim down, into the muddy habitat that they prefer. The scientists who discovered this bacteria deduced that if such bacteria live in the north, a corresponding bacteria oriented to the south should live in the Southern Hemisphere. On the basis of this theory, these scientists went to Australia, where they found a south-oriented bacteria on their first day in the field.

But the most tantalizing evidence that whales might use electromagnetic radiation from the earth's magnetic field to guide their migrations comes from the discovery of tiny grains of magnetite in the ears of the white-sided Pacific dolphin, a small cetacea similar to pilot whales. Recently scientists have started to look at geomagnetic maps of the areas where whales strand. They have found that each area has a significant magnetic anomaly located near the stranding site.

On Cape Cod an anomaly lies below the peninsula,

creating peaks of magnetism that traverse the bent hook of the Cape. As the whales travel southeast they follow these peaks. If they end up inside the bent arm of Cape Cod their electromagnetic sense, which evolved eons ago before the formation of Cape Cod, leads them to mass suicide.

A preprogrammed behavior set to respond to a specific signal is an efficient way for animals to deal with most situations. However, when the situation changes, the behavior once necessary for survival can become fatal. Despite large flexible brains, whales and humans are still held hostage to many fixed-action patterns and responses. For the whales a changed geographic situation can lead to a mass suicide. In the case of humans with an arsenal of nuclear weapons, escalating international aggression could lead to the same.

In many animals, hearing has evolved from a simple mechanical recorder of pressure to our own sense of audition. During the process it has evolved some highly sensitive detectors of other energy sources. In the future we will undoubtedly find more exotic detectors in animals and perhaps in some people who exhibit what we now call extrasensory perception.

The earth is bathed in a constant flux of energy from the sun and other stars. Most of this energy can't be detected by animals. Short wave-length energy is too powerful. It disrupts molecular bonds and mutates genes. Long-wave energy is too weak for animals to perceive. It is only between these two extremes that we find energy capable of slightly altering molecules and stimulating cells. Light waves and sound waves can do this. Light waves alter protein molecules in our eyes and sound waves stimulate hair cells in our ears. These sources of energy can be converted to electrical pulses that stimulate our brains.

In the process of evolving a brain capable of perceiving input from these environmental energy sources, we have developed a highly complex organ with billions of inter-connected neurons that communicate amongst them-selves, often without input from the outer world. All this "chatter" results in dreams, fantasies, thoughts, ideas, and language—the seeds and driving forces of civilization. The next time you use those critical little hair cells to learn about a new idea, to discuss an emerging theory, or sim-ply to listen to a Brahms concerto, remember their won-derful complexity. And remember that the times when these cells are best used is when we pass on our love for the world to a child.

18

Genes and Generators, or
How the Lobster
Evolved His Posture

Animals, whether humpback whales or horseshoe crabs, sea hares or humans, follow many patterns of behavior known as fixed-action patterns. When a squid is frightened, an electrical impulse is generated in its giant axon, contracting the mantle and jet-propelling the squid backward. When a nesting male herring gull approaches another male at the edge of his territory, the two perform a series of ritualized aggressive and appeasement behaviors. When a young man catches a woman's eye and she lifts her eyebrows almost imperceptibly and then turns her head to the side, they too are displaying fixed-action patterns.

Even if the man and woman are two highly trained lawyers with a firm belief in free will, they still act unconsciously through fixed-action patterns. Although they may be discussing some abstract principle of international law, there will be components of their behavior which are highly stereotyped, fixed-action patterns triggered by the gestures, stance, and behavior of their associate.

Despite the lofty nature of their rational conversation there is another, entirely different conversation going on. It is a highly charged, information-laden conversation about emotions, sex, and feelings. The same phenomenon would occur between two male, heterosexual physicists although their underlying conversation would have more to do with dominance, appeasement, aggression, and male bonding rather than with sex and romance.

Beneath the surface, there is yet another, far more intimate conversation occuring. This is a conversation between neurons, axons, and synapses. The same is true for other species. The posture assumed by the mating male lobster is generated by a group of six discrete neurons connected by synapses. This tiny postural circuit is like a microchip in a computer.

But unlike the microchip, the lobster's living computer is bathed in a slightly saline solution. It is this solution that gives biological computers such an advantage over the clumsy mechanical variety.

Though it may be somewhat slower to respond, the lobster's wet computer relies on a far more sophisticated electrochemical language than does our mechanical invention. Chemical neurotransmitters like serotonin, dopamine, and octopamine act on circuits of neurons to fine tune the lobster's behavior.

It is during the development of an animal's neuronal circuits and during the production of neurotransmitters that evolution can significantly alter behavior. When an animal, whether a lobster or human, is growing, it over-produces neurons. During development these neurons migrate, become isolated, and segregated into the micro-circuits responsible for behavior. If this parceling process does not occur properly, if a gene does not produce one of the many enzymes which regulate the growth and behavior of developing neurons, then the microcircuit will be faulty. This could lead to behavioral problems such as dyslexia, overly aggressive behavior or even an altered courtship pattern.

During the development of neurotransmitters, DNA in the nucleus of brain cells instructs the ribosomes to pro-duce large precursor protein molecules from which the smaller neurotransmitters are derived. Enzymes con-trolled by specific genes cleave the precursor molecules at precise locations. These discrete pieces of the precur-sor molecule become the neurotransmitters that modu-late the microcircuits in specific ways. Thus octopamine might excite every neuron in the six-neuron lobster-pos-ture circuit, but serotonin would excite two and inhibit four.

Such seemingly minor differences have far-reaching consequences. One mix of neurotransmitters causes the neurons to command the muscles to contract or relax in one pattern. Another mix causes them to contract a dif-ferent way. Two completely different postures will result from each mix of neurotransmitters. If the courtship pos-ture is wrong, mating might not ensue.

On the other hand, a minor change in a gene could cause a male lobster to assume a highly attractive posture, a super-releasing mechanism that would prove irresisti-ble to female lobsters. This male would sire many more offspring carrying his gene for a superior courtship dance. Thus behavior can enhance an individual's reproductive success or isolate it as effectively as any mountain range, ocean, or island.

A male animal with a striking new courtship dance has the chance of dominating a new line of genes. But this gives the females a choice of strategies. Some females, because their own neuronal microcircuits, might opt for a conservative strategy, to be discriminating, to wait for the male with the most stereotypical courting behavior, one whose offspring probably have the best chance of passing on their own genes. Among higher primates like baboons and humans, this behavior is common among females higher in the dominance hierarchy.

But other females, again because of their microcircuit make-up, might adopt a more liberal strategy. They would be less discriminating and quicker to mate with an atypical male. Such behavior represents more of a risk. If the environment is changing rapidly or the male has other attributes that make it more likely to survive and mate, then a liberal female strategy might win big. She could attach her genes to a star, a super male who will have many offspring that will also enjoy reproductive success.

There is nothing to guarantee the success or failure of either strategy. In fact, sexual selection can lead to some curious anomalies. Theoretically a female might mate with a male who has an unusual courtship dance or a particularly attractive attribute that seems to guarantee reproductive success, but which eventually proves to be that species' undoing.

The huge antlers of the male Irish elk come to mind. This attribute allowed the male to achieve greater reproductive success but at a large physiological cost. In the case of the Irish elk the cost was so severe that the elk is believed to have become vulnerable to predators due to the weight of his antlers and the species became extinct. We can see a more modern example in the peacock, whose magnificently attractive tail feathers undoubtedly aid its reproductive success, but which interfere with its ability to fly away from predators. Such are the precarious odds in the genetic lottery called evolution.

In humans the cost of fixed-action patterns is also severe. Jealous husbands kill their wives. Stepparents abuse their stepchildren at a far higher rate than do biological parents, and political leaders launch suicidal attacks against their enemies. Societal norms, laws, and risk-benefit analyses do something to control such behavior but it persists.

One of the reasons such behaviors persist is that they are often controlled by central pattern generators, circuits of neurons that respond to neurotransmitters and electrical stimuli within the brain. The microcircuits that control these behaviors evolved because they gave the individual fixed-action patterns with a proven survival value.

But one problem is that these behaviors were selected over thousands of years to respond to circumstances that might be encountered by hunter-gatherer societies. Today nuclear-age humans are still saddled with microcircuits designed for a far earlier age.

Most of the time our microcircuits serve us admirably, as they have for millenia. Sometimes, however, they cause us to behave atrociously—to kill those who have never harmed us, even those whom we love. In the nuclear age, they could cause us to exterminate our species. We have only our laws, our society, our morality, and our new cerebral cortex to control our often dangerously anachronistic fixed-action patterns. Let us hope they are up to it.

PART FOUR

WINTER

The Future

19

Winter

It is winter. Kittiwakes and shearwaters skim over choppy waves and leaden skies hang heavy in the arctic air.

An eddy meanders off the Gulf Stream. Now it becomes a swirling ring of warm water with a life of its own. The warm core ring will migrate southwestward through the cooler slope waters of the Atlantic until it reaches Cape Hatteras. Here, where the Gulf Stream shoulders the shelf, the warm core ring will be engulfed again by the Gulf Stream's mighty flow.

Like an underwater hurricane, the inexorable passage of this swirling mass of water wreaks havoc on local populations. Encompassed within its 90-mile diameter is the warm core. It contains the flora and fauna of the Sargasso Sea, entrained by the ring as it swirled off from the Gulf Stream. The tropical species are fated to live within the narrow compass of this isolated core for to be swept outside means certain death in the boreal waters of the Atlantic. Larvae of commercial fish are swept off the continetal shelf by the warm core ring. These larval fish will starve in the less productive waters of the open ocean, disrupting commercial fisheries for several years.

Buried within the steep-walled canyons of the continental shelf, millions of lobsters await the spring. During the fall they migrated here from their near-shore summering areas. The squid have also returned to the deeper waters of the continental slope, while cod and haddock root through the sediments on top of the shelf.

Just beyond a school of haddock, the female horseshoe crab has also completed her offshore journey. She has spent several days digging into the loose sediments. The cold waters will help consolidate the sand and she will live delicately entombed within their grip until spring. Without food, light, warmth, or activity she will hibernate here for the next five months.

Less than two hundred miles away the dark hulk of another underwater denizen, the remains of the nuclear submarine *Thresher,* lies hidden in the sediments. Deep within the leaden recesses of its nuclear-powered engine a dangerous clock marks the passage of time and the deterioration of its radioactive fuel, which has lain on the bottom since 1963. No one can be sure when, or if, these nuclear wastes will seep through the walls to enter the food chain.

The migrating horseshoe crabs have crawled past torn nets, lobster pots, dumping sites, and the abandoned holes of oil rigs—each a silent testimony to mankind's increasing use and abuse of these offshore waters. She has crawled among tumorous flounders sick with cancer from industrial waters, bluefish laden with PCBs, and striped bass, whose population is seriously depleted from the effects of acid rain.

Far overhead scientists study the status of these offshore fishing grounds. A thud reverberates through the hull of a Soviet trawler. The torpedolike casing of a bathythermograph has hit the outside of the vessel. On deck, a Soviet scientist hoists the bathythermograph and removes a tiny piece of frosted glass on which is etched a graph of temperature and depth. The graph will indicate the depth of the thermocline, the boundary of warm and cold water 4,500 feet below the warm core ring. He reports this to the chief scientist, who will radio the information to Woods Hole, where it will be compared with data from drifting buoys, satellites, and the research submersible *Alvin,* also tracking this wandering ring of water. In the laboratory the information will be analyzed to understand how warm core rings can mask the sonar signature of enemy submarines, or destroy an entire generation of commercial fish.

The steady throb of the engine dies away and the ship wallows in the nauseating cross swell of a far off storm. The trawl is over. On the bridge the captain, mate, and chief scientist hunch over the echo sounder, following the faint yellow and green lights of the electronic gear that is recording their passage over the ocean floor. They are trawling over rugged bottom topography, the jagged remains of ancient sand dunes, peat and forests that thrived here when the continental shelf was out of water during the last ice age.

On deck the ship is bathed in the eerie glow of floodlights. The bulging net emerges from the inky waters and is swung over the sorting table. A crewman pulls the cod-end knot and a silver cascade of codfish, haddock, hake, and herring spills over the sides and onto the deck, which is soon awash with fish slime, scales, and viscera. Scattered through the glistening pile are a few horseshoe crabs flexing and contracting their bodies in a vain attempt to escape.

A few hours later hundreds of fish will leer out of wire bushel baskets crowding the deck. Decreased pressure from their rapid ascent has caused their stomachs to push out of their mouths and their eyes bulge like the distorted beasts of an Hieronymous Bosch painting.

One by one the fish are separated by species, and individual fish are weighed and measured. The horseshoe crabs are counted, sexed, and measured across the carapace. After sorting the catch Soviet sailors grab a few crabs, for the cleaned and dried shells will make interesting curios in Kaliningrad.

The trawl is over. The catch is thrown overboard and the crew go below decks for tea. The female horseshoe crab descends through the cold green waters of the Atlantic back to the continental shelf. Here she will spend the winter, with luck, undisturbed until spring.

20

Eggs, Asters, and Urchins

It is mid-winter on the floor of the Atlantic Ocean. The horseshoe crabs rest quietly below the heavy sediments of the shelf. Squid are feeding in the cold green slope waters and lobsters have moved into the deep canyons at the edge of the continental shelf.

Everything is quiet on the ocean floor. Ninety feet of cold, heavy Atlantic water press down on the creatures of the bottom. These are the coldest days of the year, when the water temperature often falls below freezing but pressure prevents the formation of ice. Ever so gradually, toward the end of February, the temperature starts its inexorable rise. The change triggers and one of bio-medicine's great benefactors, the mid-winter spawning of the sea urchin, *Strongylocentrotus drobachiensis*.

The ocean floor is strewn with thousands of these black-

and-green animals. Two to four inches in diameter, the radially symmetrical creatures add color to the ocean floor. Although the sexes look identical, ninety percent of the urchins turn out to be female.

Suddenly a small puff appears above one of the sea urchins. It is a tiny cloud of millions of eggs. Pheromones from the release trigger neighboring urchins to shed their gametes into the frigid water. Within seconds, billions of eggs and sperm hangs over the ocean floor. This synchrony will continue throughout the development of the embryonic urchins.

Now billions of sperm are locked in furious competition. Powered by ATP, life's energy molecule, billions of thrashing flagellae drive DNA-packed sperm heads through the water. This is life's first and most momentous journey; the uniting of nucleic material from the DNA-packed sperm with the urchins' largest cell, the egg.

Soon each egg is swarming with thousands of thrashing sperm, attracted by pheromones from the waxy coating of the sea urchin egg. As soon as a single sperm encounters this coating, it triggers an abrupt and profound change in the sperm itself. A filament of material shoots out from the head of the sperm and this new membrane fuses with the egg.

Much like a magnetic card and a computer in a twenty-four hour bank teller, the egg and the sperm have recognized each other. The egg has allowed this single sperm entry and fertilization has occured.

The entry of the sperm triggers further changes. The egg secretes an enzyme that destroys the sites where other sperm have bound to its coating. Without suitable sites these supernumerary sperm are now sloughed off, effectively blocked from fusing with the egg. A single fast and fortunate sperm has won the first, most significant competition in life—that of fertilization.

Within the egg other changes are occurring. Intracellular reserves of calcium, the universal biological trigger, release a complex chain of events. For sixty seconds intense activity takes place within the single cell. It is followed by four minutes of quiet, as if the cell were resting before undertaking the next round of activity. Five minutes after fertilization a supressor protein is released and the egg can start its embryonic development.

Now, one of nature's most exquisite and curious organelles appears. Two tiny asters form on opposite poles of the cell. This starburst pattern of microtubules form a spindle, a loom upon which life will be woven.

Chromosomes from the sperm and egg collect at the juncture between these two spindles. Individual microtubules attach to the center of each chromosome and draw them toward opposite poles of the cell. The cell divides with each daughter cell containing half its genetic material from the egg and half from the sperm.

The DNA within each cell will now guide the development of the proteins and enzymes that will shape the sea urchins' morphology and behavior. It is through this elegant dance of the chromosomes that the accumulated wisdom of billions of years of evolution is passed on to future generations.

The detailed description of the development of the sea-urchin egg was the first unique contribution of American biology to science. Charles Otis Whitman, while studying in Germany in 1878, elegantly proved to his European mentors that development of the sea urchin embryo does not simply duplicate the evolution of that species as Haeckel had predicted.

Free from the shackles of this outdated theory, American biology, formerly a second-rate endeavor, continued on the road of discovery that was to lead to the vast potential of gene-splicing.

Charles Whitman went on to become director of the Marine Biological Laboratory in Woods Hole where so many of these discoveries were made. Today MBL scientists from all over the world still converge on Woods Hole to look at the development of the sea urchin embryo to discover further fundamental secrets of life.

Strongylocentrotus drobachiensis, one of the many urchins under investigation, has the dubious distinction of having the longest two-part name in biology and the perverse habit of mating during the coldest days of the year. The latter distinction makes it far more useful than the former. Most animals cannot form and maintain the microtubule spindles necessary for cell division under extremely low temperatures and at high pressures.

Today scientists investigate the development of microtubules with electron microscopes and protein separation techniques rather than the hand lenses, collecting jars, and formaldehyde of earlier "bucket biologists." In addition to making contributions to our understanding of genetics and development, the study of microtubules has led to a greater understanding of several diseases.

One type of sterility occurs in males who lack a gene for the structural protein dynein, which holds the sperm's tail microtubules together. Without this enzyme the tail of the sperm cannot thrash through the water, meaning the sperm cannot travel and fertilization cannot take place.

Another devastating disease, cystic fibrosis, attacks young children, leaving them unable to breathe. A protein factor in the blood disrupts the regulated flow of calcium ions that control the interaction of dynein and microtubules in the cilia of the lungs. The effect is so pronounced that scientists have used it to perform a simple diagnostic test. Blood serum from a child suspected of having the disease is placed on the gill of an oyster. If the cilia of the gill stop beating, the child has the dreaded disease.

Finally there are several diseases of the brain that cause progressive dementia. The most common of these is Alzheimer's disease, which is caused by knots of tubulin protein in the brain. These faulty microtubules block the flow of neurotransmitters necessary for ordered thinking and motor control.

Scientists have been hard at work trying to find the causes of these faulty microtubules. They suspect that problems with the genes responsible for the formation of the protein tubulin, as well as chemical agents like aluminum, could alter the formation of microtubules.

Now, a new, potential cause has been reported. In New Guinea there are reports of a disease, *kuru,* that is passed from tribe to tribe, although the mutually antagonistic groups involved have little interaction other than through warfare. Anthropologists have determined that the disease is transmitted through cannibalism. Warring tribes eat the brains of their victims in order to possess the courage of their enemies.

It was recently discovered that the agent responsible for this disease could be a slow virus that lacks DNA. It has been dubbed a "prion," short for infectious protein. This controversial substance may prove to be one of the causes for such diseases. If this is the case, the mechanism for its destruction might be found through research on the sea urchin, *Strongylocentrotous drobachiensis,* the cold-water breeder of the North Atlantic.

21

Coastal Cancer

As the great arc of the winter sun moves toward the vernal equinox, it warms the shallow coastal waters of the eastern seaboard. Ice breaks up and moves down the estuaries below Boston, Philadelphia, Baltimore, and Washington, D.C.

Beneath these waters life stirs once again. Winter flounder, *Pseudopleuronectes americanus,* have grown ripe with the next generation. A female heavy with roe bursts from beneath the sediments. Undulations ripple along the fins fringing her body. She glides to a stop, resting gently on the single pectoral fin below her body. From this slight elevation her bulging eyes can survey the estuary floor.

She detects a slight movement. The head of a nereanthid worm emerges from the seafloor and tentatively explores the surrounding sediments. With a single fluid motion the flounder pounces on her prey. One of the only large predators still active in the winter estuary, she has benefited from the lack of competition to grow sleek and fat, ready for spawning.

But this year she is slowed by a painful tumor growing within her liver. She swims toward a pipe where raw sewerage boils out into the estuary. Other flounder have congregated here, absorbing the carcinogenic broth.

Now she is followed by several males who sense that her time of spawning is at hand. She scrapes her swollen belly across the bottom. One egg emerges, then another, then an explosion of thousands of sticky translucent eggs spew from her swollen vent. They fall to the estuary floor, adhering to pebbles, rocks, an empty bottle, a pile of wastes dumped here several years ago. The males crowd around her, emitting great clouds of sperm that settle on the sticky eggs. Soon the estuary will be replete with jellyfish whose development has been timed to exploit the rich harvest of larval flounder.

There are several species of flounder that inhabit the East Coast of North America. Through time they have evolved distinct behaviors to exploit different niches of the area. Some live offshore in cold deep water year-round, others migrate. Some have small mouths for feeding on invertebrates, others have evolved large mouths for feeding on bait fish.

But inside the eggs, the embryonic flounder are also being bathed in this carcinogenic chemical soup. It is the beginning of a molecular chain reaction that will eventually lead to cancer by the time many of the fish become adults.

Unfortunately it is the constellation of behaviors sur-rounding winter spawning that leaves the winter floun-der at such high risk. Originally they evolved wintertime spawning; heavy, sticky eggs; and short migrations to keep them in the estuaries during winter when other preda-tory fish had moved offshore. These behaviors cause them to live in small, local populations inside estuaries. For-merly this behavior gave them a great advantage over other fish. Today, however, this repertoire of behaviors can concentrate winter flounder in urban estuaries near sources of carcinogenic pollution.

The sequence of molecular changes leading to cancer proceeds in several distinct steps. One carcinogen must gain entry to a cell's DNA where it can initiate a mutation. Later a second type of carcinogen must be present to pro-mote the proliferation of the newly mutated cells.

Flounder, like humans, have their own cellular defense systems to stop the progression of cancer. This is why it often takes years of exposure to both intiator and pro-motor carcinogens for enough mutations to accumulate to cause cancer. This process suggests a strategy for sub-verting the process: administer a chemical substance that blocks the cellular receptor sites for promotor carcino-gens. Some researchers are looking at diet as a way to prevent cancer in humans. Orange and green vegetables, like carrots and broccoli, contain beta-carotene, which seems to have the ability to block a cell's receptor sites for some promotor carcinogens.

If both the initiator and the promotor carcinogens are present, the flounder might develop a benign tumor, benign only in that it is not spreading wildly. It is deadly in that it can still cause death by blocking blood vessels or the trachea. If the tumor "turns on" and becomes malignant, it behaves much like an avaricious empire run amuck. Through metastasis, individual cells regain their embryonic ability to generate enzymes able to dissolve blood vessel walls. Once inside the blood vessels, the cells are carried to new locations where they establish more tumors. Wedges of malignant cells push through living tissue, consuming all available nutrients. The wedges look so much like crab claws that Hippocrates named the disease after the Greek word for crab (karkinos). Today the term survives as cancer, Latin for crab.

The cancer process is so much like that of an expanding empire that one cannot help but use the analogy. However, the analogy goes further. The new colonies proliferate and have a life of their own. They even attain an immortality of sorts. In laboratories throughout the world there are petri dishes that contain isolated colonies of malignant tumors thriving long after they have killed the humans that first gave them life.

Coastal industrial cities are slowly grappling with this plague of offshore cancer. It is a problem that can only be met by expending millions of dollars, overhauling many governmental bodies, and changing the way we deal with urban wastes. In Boston, for instance, the citizens are locked in a drawn out political battle. It is a colorful struggle in the way that only Boston politics can be colorful and passionate only in the way that politics, the Red Sox, and the Celtics can generate Boston passion.

In 1983 the federal court mandated that no more construction permits could be issued in Boston until Massachusetts came up with a way to clean Boston Harbor. After considerable Sturm und Drang, the state legislature was able to create the Massachusetts Water Resources Authority.

The authority must now devise a plan to build new sewage treatment plants that will meet federal water quality requirements. Every day spent wrestling with these problems means another 164 million gallons of partially treated sewage spill into Boston Harbor and an epidemic of cancers multiply in its waters.

Because they have evolved behaviors that exploit the wintertime estuaries, winter flounder have become allies in mankind's efforts to stop the spread of offshore cancer. Like canaries used by early miners to detect the build-up of poisonous gas in tunnels and mine shafts, flounder have become one of our newest "indicator species" to detect carcinogens in coastal waters.

Indicator species have played an important role in defining, as well as symbolizing, many environmental problems. During the 1960s Rachel Carson warned of the danger of a "silent spring" caused by the use of DDT and other pesticides. Eagles, ospreys, terns, and pelicans became the symbols of the problem as their numbers dwindled because of egg failure. DDT, concentrated in their diet of fish, was affecting their ability to deposit sufficient calcium in their eggs.

Lobstermen have long known that a kerosene-soaked brick placed in a lobster pot would attract lobsters. Because a small amount of oil will alter their feeding and courtship behavior, lobsters have been used experimentally to measure the effects of oil spills.

Mussels are being used in a world-wide marine-pollution alert program called "Mussel Watch." The animals make ideal standard sampling devices because they have a world-wide distribution and they feed by filtering gallons of sea water every day. By sampling their tissues, researchers can compare pollution levels from all over the world.

Striped bass, shad, and herring, marine species that breed in brackish-water estuaries, are declining because of acid rain. They have become indicator species for brackish water, while winter flounder, along with salmon, sole, tomcod and soft-shelled clams have become indicator species for coastal waters.

Someday soon, species like flounder may be used instead of laboratory rats to test the thousands of new chemicals introduced onto the market every year. We would then be informed of the chemicals' carcinogenic properties before they enter the environment.

For now, we are left with these marine animals, once considered immune to the effects of man, as new indicator species, symbols of the marine "silent spring" that is rapidly engulfing our coastal waters.

22

Seals, Diving, and The Human Heart

The tide is rising on Grassy Island, a random string of rocks in Woods Hole Harbor. A small pod of harbor seals, *Phoca vitulina concolor,* bask in the early morning sun. A young male casually rests his flipper upon the body of a dozing female. Glowering, she pulls away. He follows. She slips into the water. He dives after. Splashes abound as the two roll, leap, and chase each other through the shoal water. Excitement spreads through the incorrigibly polygamous animals. Soon the entire pod of young seals frolic in the water.

The increasing daylight has initiated this mating behavior and will soon trigger their migratory behavior. By early spring these seals will be joined by others from as far south as Long Island. Together they will migrate to their summer rookeries in Maine and Canada. After mating, delayed implantation of the developing embryo will bring them into synchrony with the calendar year. This is but one of the adaptations they have evolved to cope with their curious situation, animals suspended in transition between a terrestrial and a marine existence.

But now the season is early. Ardor gives way to hunger and the animals turn their attention to feeding. The young female seal takes a breath and dives straight for the bottom. Upon entering the water, an exquisitely orchestrated repertoire of changes occur. Her heart rate plummets, blood vessels constrict, and her lungs start to collapse. Less blood flows to non-vital organs like the kidneys, but the blood flow to the brain remains constant. The physiological changes have been so well coordinated that there is little change in overall blood pressure.

On the bottom of the channel the rapidly flowing water is clear and cold. Starfish creep over mussel colonies and kelp fronds sway in the incoming tide. The seal levels off and cruises over the bottom. Faint light is picked up by her large bulbous eyes, which contain more dim-light sensitive rods than color-sensitive cones. Now she has spent several minutes on the bottom. She is still drawing oxygen from the hemoglobin molecules in her blood and myoglobin in her muscles, of which she has far greater quantities than non-diving terrestrial animals.

If an emergency should occur and she must stay underwater longer, her muscles can switch to anaerobic respiration or breathing without oxygen. In humans, lack of oxygen would lead to brain death within minutes. In seals this little understood back-up system gives an extra margin for safety. This time it is not necessary. A flounder darts from a patch of seaweed. The seal scoops it up and

sweeps for the surface, emerging with the flapping flounder held crosswise in her jaws. The dive has taken eight
minutes. Her heart rate resumes its normal, faster terrestrial pace.

The most dramatic response seen in diving animals is
bradycardia, the marked slowing of the heart. It was first
described by the French scientist Paul Bert in the 1870s.
Bert strapped two ducks to a board, plunged the unfortunate animals into cold water, and recorded their heart-
beats. He described the changes as part of the "diving
reflex."

Every human has experienced a similar response during birth, and many have relived it as parents in the deliv-
ery room. Primed with all the latest theories and
techniques, the expectant parents go through a battery of
breathing exercises. They marvel at the nurses' ability to
determine their baby's sex by its heart beat. Then suddenly something goes wrong. The fetal monitor flashes,
the baby's heart rate has dropped. This is when most rea-
sonably neurotic parents forget everything they have been
told and expect the worst. But this is simply the "diving"
response—the fetus is starting the fateful plunge into the
terrestrial world.

The diving response is also the reason that people, particularly children, can recover after a long submersion in
cold water; in order to protect the brain, the body slows
the heart and constricts blood flow to non-essential organs.
These changes have implications for humans in the study
of heart attacks, diabetes, stress metabolism as experienced by marathon runners, and in sudden infant death
syndrome.

Bert described the changes that occur during diving as

an automatic reflex. Other researchers felt that it could be an artifact of his technique. Simply put, his ducks could have been scared to death. Like an accused Salem witch strapped to a dunking chair, the ducks didn't know if or when they would be brought to the surface.

Modern researchers felt they should study the response in unrestrained animals. One way of doing this was to raise diving animals as pets. Dr. John Kanwisher, a researcher at the Woods Hole Oceanographic Institution, did just that.

Dr. Kanwisher collected the eggs of cormorants, diving birds known for their prowess at catching fish underwater. The chicks imprinted on his wife and would come whenever Dr. Kanwisher whistled. One of the unfortunate consequences of this training occurred at a lavish cocktail party. Dr. Kanwisher, wishing to show off his cormorants to the wife of one of his supporters, stepped to the window and whistled. A cormorant flew in, landed on the table, and proceeded to stomp through the hors d'oeuvres, knock over drinks, and evacuate great quantities of guano in its excitement at being included.

Other researchers chose to study mammals. But large diving marine mammals make notoriously poor research subjects. In the laboratory they are ungainly, intractable, and always hungry. In the wild they are difficult to find and complicated to observe because they spend most of their time underwater. Most importantly, as mammals they are frustratingly complex. This is the reason that most scientists searching for fundamental truths about life stick to simple tractable organisms like *E coli* bacteria, fruit flies, and horseshoe crabs. Such scientists have even built up a certain conceit implying that serious scientists stay in the laboratory while frivolous animal watchers venture outside.

However, generation after generation, handfuls of stubborn investigators have gone into the field with little more than notebooks, binoculars, and a keen sense of observation. Each time they do, they emerge with new insights that challenge accepted theories, occasionally win Nobel prizes, and keep laboratory scientists busy for years.

Many of the more celebrated field studies of exotic marine animals like whales attract a great deal of attention, but they are often quite primitive; the researchers are simply trying to get counts and identify their research animals. But one less well-known, less heralded but intriguing study of seals has emerged: the study of the Weddell seal, which lives below the fast ice surrounding coastal Antarctica.

Weddell seals, *Leptonychotes weddelli,* evolved from a northern stock that settled on the temperate Antarctic continent before it moved to its present location. As the continent moved south, its climate grew more harsh and the Weddell seals evolved behaviors to deal with these new extreme conditions. Today it is the only mammal living this far south and the only large predator able to fully exploit the year-round abundance of Antarctica.

For the Weddell seal every dive could be fatal but every dive also holds the opportunity for evolution. The drastic changes in the functioning of its heart, lungs, and blood hold profound implications for human health.

The modern world of the Weddell seal is forbidding and frightening. The long winter night envelops Antarctica for three months. The only light comes from the moon, the stars, and the brilliant shimmering hues of the aurora australis. By June, the beginning of winter, the killer and minke whales have gone. The Adelie and Emperor penguins, the leopard and crabeater seals have all migrated north, only the Weddell seals and a few human researchers remain. . . .

An adult male seal prepares to dive. He takes a breath and plunges into a long dark tunnel of ice. He has maintained this breathing hole through constant attention. When the ice was thin he could bash it open with the top of his head. As the ice thickened, he would ream it open with specially evolved canine and incisor teeth. Now, by late winter, the tunnel is over forty-five feet deep, and only wide enough for the seal's body. The tunnel passes through six feet of snow lying on top of twelve feet of fast ice. Under the fast ice, thirty more feet of loose platelet ice threaten to break off and clog the tunnel.

Below the tunnel the water is crystal-clear but dark. Foot-long crystals of ice grow on the ocean floor. These will eventually break off and float to the undersurface of the ice. The seal reaches one hundred fifty feet and cruises at this depth. It passes a foot-long icefish, *Chaenocephalus aceratus*. Almost transparent, the fish lacks any red blood cells (hemoglobin or myoglobin). Its gills are white and its heart pale yellow. Instead of blood pigments it contains a little understood natural antifreeze that allows it to be frozen solid in ice yet still survive.

Normally the seal would catch this fish but this is an exploratory dive: he is searching for another breathing hole. He cannot afford to surface and swims on through the dark. He has now traveled 3.5 miles from his breathing hole and has been underwater for over twenty minutes. Since he cannot locate another breathing hole, he must return unerringly toward his own. Confusion in these dark featureless depths could prove fatal.

The seal emits a cacophony of tweets, twitters, chirps, and squeals to help it identify abnormalities in the ice overhead. He has not been this way before and must "sound" his way back. His lungs have collapsed from the depths, his heart rate has dropped, and he has switched to anaerobic respiration, metabolizing fat cells without using oxygen.

Finally the seal locates its tunnel entrance but a smaller seal is near. The two animals rush toward each other. The adult seal must fight off this intruder before its oxygen runs out. If the second seal swims into the tunnel first it could block the male's access to air.

The first seal expels a large air bubble that floats to the undersurface of ice. He rushes back to challenge the intruder, who flees. The male returns to the shallow air bubble he left below the ice. Some carbon dioxide has seeped out of the bubble. The seal rebreathes the remaining oxygen and rushes back toward the tunnel.

He surfaces rapidly toward the dim light. His head breaks the surface and he expels a great gasp. Two researchers snap into action. They have been waiting for the seal's return in a small warm laboratory placed over his breathing hole. The seal lets them slip off a harness that holds a small depth gauge. They measure the expelled air, and disconnect some electrodes that recorded its heartbeat.

Fortunately for scientists, the habitat of wild Weddell seals makes them ideal research animals. Living in the Antarctic, they have developed no fear of mankind. During the long antarctic winter they might even enjoy the companionship of researchers, despite their curious penchant for light, heat, and a strange assortment of instruments.

Occasionally the seal widens his breathing hole too vigorously, threatening to undermine the researchers' laboratory, and sometimes differences of opinion occur. Researchers who sneak up on a nursing mother and push aside her pup in order to take a milk sample from the mother's nipple are often sent sprawling by an admonishing whack of her flipper.

Scientists can observe seals from underwater chambers and even dive through the seals' long breathing tunnels, but this is not often done. Divers relate frightening stories of becoming disoriented under the ice and fearful that a seal would beat them back to the tunnel, sealing their fate to a slow death below the frigid antarctica ice.

It is situations like these that give fieldworkers a visceral appreciation of the feats their research animals can perform. They have also learned that animals' responses to diving are not just automatic reflexes but are often voluntary responses to different situations. If a Weddell seal expects to make a fishing dive of less than twenty minutes, minor physiological changes occur. If the seal expects to make a longer exploratory dive, then more drastic changes take place.

Because seals can voluntarily control their heartbeat, they may provide clues to preventing heart attacks in humans. Heart and kidney transplants may become more successful if we can find out how seal's organs can withstand long periods without oxygen, and changes in blood sugar during diving may prove to have positive implications for diabetics. The constellation of physiological changes that occur in the diving response also hold promise for preventing sudden infant death syndrome.

At the same time researchers have learned about these human implications, they have learned more about an animal which is one of the wonders of the world, a creation equal in craftsmanship to the pyramids of Egypt or the works of Beethoven.

It is fitting that white-coated harp-seal pups have become one symbol for animal rights. Marine mammals, like many of the so-called "lower" research animals, are furthering our knowledge of human health and behavior. They deserve our gratitude for the human lives they have helped save, and the knowledge they have helped us accumulate. Most of all we should feel reverence for sharing the planet with such exquisite works of nature.

EPILOGUE

~~~~~~

THE FUTURE OF HORSESHOE CRABS

It is 8:30 A.M. A hush descends on the group of neurosurgical residents hovering over the operating table. The instructor makes the first incision. Whispers are traded back and forth as a spurt of blue blood oozes out of the wound.

Normally, these surgeons would be clustering around the operating arenas of some of the best American hospitals. Today they crowd around a simpler table, their patient a horseshoe crab. These are graduate students, neurosurgical residents, and medical school faculty who are attending the neurobiology course at the MBL in Woods Hole. They have taken two weeks from their regular studies in order to attend this course, which will give them an update on the latest developments in neurobiology.

Outside, the cold winds of a Cape Cod nor'easter stir

whitecaps on Eel Pond and swirl eddies of snow against the windows. Meanwhile, the instructor steadies a horse-shoe crab and drills a small hole next to its compound eye. He inserts a hacksaw into the hole, cuts a circular piece out of the carapace, and lifts the eye out of the crab so that only the long white optic nerve connects it to the animal. With one quick snip, he severs this bundle of fibers so that an inch of optic nerve dangles from the excised eye.

The eye, still embedded in the carapace, is given to my group of students. We are a female medical student from Texas, an ophthalmologist from Cincinnati, and a horse-shoe-crab farmer from Cape Cod.

We take the excised eye into a small darkened laboratory where it is mounted in a tiny vice so that the optic nerve trails into a shallow bath of saline solution. The ophthalmologist from Cincinnati combs out the optic nerve, separating the individual fibers that connect to one of over sixty photoreceptor cells in the compound eye. We then connect a single fiber to a suction electrode. Faint static is heard until the machine is adjusted, then the steady rhythm of electrical impulses comes over the speaker and the monitor displays a series of pulsating green spikes.

We scan a pinpoint of light across the compound eye until it falls on the single cell to which the fiber is con-nected. By shining the light on an adjacent lens, the elec-trical impulse is muted: this is lateral inhibition. We have recreated Dr. H. Keffer Hartline's pioneering experiment elucidating one of the general underlying properties of sensory systems.

I have come to Woods Hole to see what the future holds

for horseshoe crabs. In doing so, I am witnessing a cross section of research on the cutting edge of marine biology at three of the world's most prestigious scientific instruc-tions.

The Marine Biological Laboratory continues its hundred-year tradition of advancing basic research and passing the results on to clinicians. The Woods Hole Oceanographic Institution, a center for "blue-water oceanography," con-tinues to unravel the secrets of the deep sea. The institu-tion's horseshoe crab research is ancillary to this main task. Indeed, it was Dr. Watson's interest in deep-sea marine bacteria that led him to *Limulus* amoebocyte lysate, an offshoot of his primary research and one he originally intended to follow for only a few weeks.

I have come to Woods Hole specifically to seek out the National Marine Fisheries Service, which has been con-ducting research that may determine the future of horse-shoe crabs. During my earlier work at Pleasant Bay we started to tag the animals. Our first tags were balloons that we tied to the crab's tails. Throughout one summer, sailors were mystified by thousands of brightly colored balloons gaily crisscrossing the bay, seemingly under their own power. Later we advanced to Day-Glo paint, and finally plastic tags. From the tag returns we started to get a pic-ture of the horseshoe crab population in Pleasant Bay.

Pleasant Bay is the main collecting area for Associates of Cape Cod, the largest producer of *Limulus* amoebo-cyte lysate. From April through October the company bleeds five hundred crabs a day and would like to increase production to one thousand a day and eighty thousand a summer. This compares to 50,000 cats used per year for research. The bay presently has a population of forty thousand large female crabs, preferred by the bleeders because of their size. In 1982 close to half of these crabs were bled for lysate.

We made the further unsettling discovery that often over fifty percent of the bled crabs die before being returned to the water. Autopsies reveal that up to two weeks after being returned, even more crabs die because of blood clots in their hearts.

As local populations have been depleted, bleeders have had to go farther and farther afield to get their crabs. In 1985, Associates of Cape Cod got half its crabs from Narragansett Bay while a New Jersey company got all its crabs from South Carolina.

Just prior to the Civil War a similar situation arose in the oyster industry. When fishermen depleted the oyster beds of Cape Cod they started to move south, first to Narragansett Bay, then to Long Island, until they finally settled into what became known as the Virginia Trade. Cape Cod fishermen would anchor in Chesapeake Bay off of Maryland and Virginia, dredge up oysters, and transplant them to Cotuit or Wellfleet. After a few weeks these transplants would pick up the flavor of the Cape Cod waters and could be sold as Cotuits or Wellfleets. But this was accomplished at great cost to the consumer and high mortality among the oysters. The practice ended with the Civil War, but it is not recorded whether this form of Yankee ingenuity helped lead to it.

To avoid this sort of problem, the National Marine Fisheries Service has undertaken a census of East Coast horseshoe crabs.

Since its inception, the National Marine Fisheries Service has expanded the scope of its activities as new fishing industries have developed. Its scientists now study deep-water lobsters in response to the development of lobster fishing in the deep-water canyons on the edge of the continental shelf. Squid populations are being surveyed because of the new export markets have developed since the United States enacted the two hundred-mile fisheries limit, which prevents foreign vessels from fishing within two hundred miles of the United States coast. Horseshoe crabs are being censused because of their newfound value to the pharmaceutical industry.

The Fisheries Service conducts offshore research cruises during the autumn and winter. Its researchers are finding that the bulk of the East Coast horseshoe crabs winter offshore between Atlantic City, New Jersey, and Chincoteague, Virginia. Few are found north of Long Island or south of Cape Hatteras.

The service is also conducting a survey of scientists and commercial fishermen to determine the past and present uses of horseshoe crabs. During the 1930s, for instance, horseshoe crabs were used in poultry feed and fertilizer. Up to one hundred thousand crabs were collected daily and ground up in plants along Delaware Bay. Even place names reflect this practice: Slaughter Beach on the Delaware shore was one of the prime collecting areas for this grim occupation.

But the foul odors of the decomposing crabs and the rank taste they lent to chickens soon brought an end to this industry. During the 1950s horseshoe crabs were considered serious shellfish predators. Bounties were levied on horseshoe crabs in Massachusetts and scores of Cape Cod children made summer money by turning in the crab tails for five cents a piece.

THE YEAR OF THE CRAB

During the last few decades horseshoe crabs have become the preferred bait for eel and conch fishermen. The Food and Drug Administration estimates that up to a million of the large egg-laden female horseshoe crabs are used for this purpose.

The Food and Drug Administration, which also sets regulations for the use of horseshoe crabs in the *Limulus* lysate industry, estimates that over a million crabs a year are used for bleeding purposes. The FDA requires that the crabs be returned to the water within seventy-two hours of having been bled. This regulation was based on research conducted in Florida where the crabs were bled and released in the field.

In Pleasant Bay we found that under industrial conditions, up to fifty percent of the crabs often died before being returned to the water. This usually occurred when something went wrong at the bleeding company and hundreds of crabs were left out in the sun after being bled.

Improvements are being made. The Food and Drug Administration is considering changing its regulation so that the crabs must be returned to the water within twenty-four hours of having been bled. Many companies have started to use mobile laboratories for bleeding. The lab is simply driven to an estuary where the crabs can be captured, bled, and returned to the waters immediately after bleeding.

If North American horseshoe crabs are to continue helping mankind, we should pay attention to what happened in Japan, where horseshoe crabs used to be used for bait, pets, and food. On almost any street vendors sold the tethered animals for pets. Restaurants served horseshoe crab eggs, a sort of poor man's caviar. The Japanese government finally had to step in to declare the *Kabuto-gami* endangered, and it is now a protected animal in Japan. In this country, there has recently been recognition of the importance of horseshoe crabs. In May 1986, the governors of New Jersey and Delaware declared a strip of beach where the horseshoe crab mate as a sanctuary to protect the crab eggs, which are a major food source of migratory birds.

Throughout the 4.6 billion years of its existence, our planet has accumulated an impressive inventory of living plants and animals, a diverse gene pool that should be protected and passed on as our most valuable asset to future generations. There is probably no more compelling example of the importance of maintaining the world's gene pool than that of the horseshoe crab.

It is the elegant simplicity of its primitive system that has made the horseshoe crab so well adapted to its shallow-water environment. Had horseshoe crabs altered over time or become extinct, mankind would have lost the genes for a unique biological organism which is now saving human lives because of this primitive heritage. With luck, research, and a helping hand, horseshoe crabs should be around for another million years or so. Can the same be said of man?

BIBLIOGRAPHY

PREFACE: TO WOODS HOLE

In the summer the village of Woods Hole has the greatest concentration of scientists in the world. There is little wonder that its rich history has been so well documented. For an attractive, readable account see Mary Lou Smith, *Woods Hole Reflections* (Woods Hole: Woods Hole Historical Collections, 1983).

I probably spent more time verifying Tallulah Bankhead's quote about Woods Hole than any other fact in this book. Virtually every summer resort up and down the eastern seaboard claims the quote as theirs. In Maine I heard that she said, "You can take Prout's Neck and shove it up Woods Hole." On Martha's Vineyard the story was told about Gay Head. To the best of my knowledge, she first uttered the phrase while being escorted off of Nantucket by the board of selectmen because of her all-night parties. In any event, my research leads me to believe that once she recognized a good line she milked it for all it was worth. See Kieran Tunney, *Tallulah, Darling of the Gods: An Intimate Portrait* (New York: E. P. Dutton, 1973).

1. SPRING ARRIVES ON THE CONTINENTAL SHELF

There are several good textbooks on marine biology, although most concentrate on West Coast fauna. For one of the best about the East Coast see Matthew Lerman, *Marine Biology: Environment, Diversity, and Ecology,* (Menlo Park: Benjamin Cummings Press, 1986).

For a good description of the continental shelf communities see Marvin Grosslein and Thomas Azarowitz, *Fish Distribution* (New York: Sea Grant, 1982).

2. THE PALEOZOIC CRUCIBLE

Our understanding of the early evolution of life is constantly changing. For one of the most readable descriptions see Lynn Margulis, *Early Life* (Boston: Science Books International, 1982). The foremost proponent of punctuated equilibrium is Stephen Jay Gould. See his book, *The Panda's Thumb* (New York: W. W. Norton and Co., Inc., 1980).

3. THE NIGHT OF THE SQUID

Octopus and squid have long held a fascination for zoologists and epicures. For a good description of their natural history see Jacques-Yves Cousteau, *Octopus and Squid* (London: Cassell, 1973).

For a description of the squid fisheries see Ann Lange, *The Status of Squid Stocks* (Washington, D.C.: National Marine Fisheries Service, 1986).

For a description of the squid's use in neurobiology see Arnold et al., *"Loligo pealei"* (MBL Laboratory Guide: 1974).

4. THE LOBSTER POSTURE

For the best description of lobster courtship see Lewis Carroll, *Alice's Adventures in Wonderland* (New York: Signet Classics of the New American Library).

More recently, lobster courtship has been described at Harvard Medical School and the MBL. Doctors Ed Kravitz, Jella Atema, Bruce Johnson, and Paola Borroni were particularly helpful to me in this chapter. See Jella Atema, "Smelling and Tasting Underwater," *Oceanus* (Fall 1980).

5. AT AN ANCIENT ORGY

Recently horseshoe crab mating has received wider public attention. Delaware and New Jersey have established a joint shoreline sanctuary to protect both horseshoe crabs and the migratory birds that depend on their eggs. See Erik Eckholm, "Spring Rite of Gluttony Fattens Birds for Journey," *The New York Times* (May 20, 1986). Also see Jack and Anne Rudloe, "The Changeless Horseshoe Crab," *National Geographic* (April 1981). For perhaps the best description of horseshoe crab mating see William Sargent, *Shallow Waters: A Year on Cape Cod's Pleasant Bay* (Boston: Houghton Mifflin, 1981).

6. BIRTH AND THE GENETIC SEX SWITCH

For the development of the horseshoe crab egg see Kathleen French, "Biomedical Applications of the Horseshoe Crab," in *Progress in Clinical and Biological Research*, vol. 29 (New York: Alan B. Liss Co., 1979). Irven DeVore and E. O. Wilson of Harvard University have been very influential in explaining the evolutionary significance of males. For an excellent discussion of human nature and the role of males in evolution see Melvin Konner, *The Tangled Wing* (New York: Holt, Rinehart and Winston, 1982).

7. THE THORN IN THE STAR, THE GOUT IN THE FISH

To be absolutely *au courant* in the zoological set, one should refer to starfish as sea stars. I believe that no one is going to mistake an echinoderm for a fish and thus stubbornly persist in using the old-fashioned term starfish. The best source for a description of Metchnikoff's work is in his own writing; see Elie Metchnikoff, *Immunity in Infectious Diseases* (New York: Johnson Reprint Company, 1971).

Doctors Gerald Weissmann and Sylvia Hoffstein are the best authorities on their work with dogfish and gout; see Gerald Weissmann, "Metchnikoff Revisited," *Oceanus* (1976). Also see Sylvia Hoffstein, "Mechanisms of Lysomal Enzyme Release," *Arthritis and Rheumatism* (1964).

For a discussion of early medical practices see Lewis Thomas, *The Youngest Science: Notes of a Medicine Watcher* (New York: Viking Press, 1983). The bible for fish scientists is still Henry B. Bigelow and Schroeder, "Fishes of the Gulf of Maine," *Fisheries Bulletin,* (Dept. of the Interior, 1953).

8. ECDYSIS: HORSESHOE CRAB ADOLESCENCE

The best description of a marine biologist is John Steinbeck's loving portrayal of his close friend and mentor Ed Ricketts, the anonymous "Doc" in Cannery Row; see John Steinbeck, *Log from the Sea of Cortez* (New York: Viking Press 1951). Louis Liebowitz, who kindly read my manuscript, has done most of the research on the infection of horseshoe crab gills by *Bdelloura candida* flatworms.

9. PENIKESE AND AGASSIZ: WOODS HOLE BEGINNINGS

In addition to the sources cited for the preface, see Lewis Thomas, *Lives of a Cell: Notes of a Biology Watcher* (New York: Viking Press, 1974). Also see Gerald Weissmann, *Woods Hole Cantata: Essays on Science and Society* (New York: Dodd, Mead & Company, 1985).

10. SERENDIPITY: THE DISCOVERY

Serendip was first mentioned in the Koran as the island where Adam and Eve fled after being expelled from the Garden of Eden. It's present usage was first introduced by Dr. Walter Cannon, physiologist at Harvard Medical School.

For a discussion of Dr. Gram's work with pyrogens see Lewis Thomas, *The Youngest Science* and "The Physiological Disturbance Produced by Endotoxins," *Annual Review of Physiology* (1954). Mrs. F. B. Bang was very helpful in verifying events surrounding Dr. Bang's pioneering work; also see F. B. Bang, "A Bacterial Disease of *Limulus polyphemus," Bulletin of John's Hopkins Hospital* (1956).

To learn more about horseshoe crab blood see Elias Cohen, "Biomedical Applications of the Horseshoe Crab *(Limulidae)," in Progress in Clinical and Biological Research,* vol. 29 (New York: Alan R. Liss, 1979). This book covers some of what we know about the biology, ecology, and biomedical use of horseshoe crabs.

11. GOLD EXTRACTED FROM BLUE BLOOD

Much of this discussion of the *Limulus* lysate industry comes from personal observation and discussions with Carl Schuster and officials of the Food and Drug Administration. Also see Peggy Thompson, "Value Extracted from a Nuisance," *Smithsonian* Magazine, (April 1974).

12. THE MIGRATION

Animal migration has interested scientists for a long time. There are several recent books that describe what is presently known about animal's migratory behavior; see K. Schmidt-Koening and W. T. Keeton, *Animal Migration, Navigation, and Homing* (Berlin: Springer Verlag

1978.) Also see Carthy, *Animal Navigation* (New York: Charles Scribner & Sons, 1956).

13. A SHORT COURSE IN NEUROBIOLOGY

Neurobiology is presently the most challenging, fastest-paced field in biomedical research. As new information is obtained, new theories are developed. For one of the most up-to-date, readable, graduate-level textbooks see Gordon Shepherd, *Neurobiology* (New York: Oxford University Press, 1983). For some of the classics in the field see K. C. Cole, *Membranes, Ions, and Impulses* (Berkley: University of California Press, 1968 and William Adelman, *Biophysics and Physiology of Excitable Membranes* (New York: Van Nostrand Reinhold, 1971). For references to use of red-tide toxins in espionage see "CIA Report," *Congressional Record* 1975–1976.

14. THE AGE OF AEQUOREA: THE JELLYFISH THAT LIGHTS UP THE HUMAN BRAIN

For a comprehensive review of jellyfish see Rees, *The Cnidaria and Their Evolution* (London: Academic Press 1966). For a description of the use of aequorin see Woodward Hastings, "Bioluminescence," *Annual Review of*

Biochemistry (1968) and "Bioluminescence," *Oceanus* (1976). Also see Zahl, "Nature's Night Lights," *National Geographic Magazine,* 1971.)

15. THE BRAIN, THE NERVE, AND THE SQUID

In addition to Shepherd's textbook on neurobiology, there are several books and articles about squid, neurons, and learning; see William Adelman, "The Squid's Giant Axon," *Oceanus* (1976). For a discussion of the brain by one of the pioneers see J. Z. Young, *Programs of the Brain* (Oxford: Oxford University Press, 1978. Robert Barlow helped with personal reminiscences of H. Keffer Hartline and J. Z. Young's discovery of *Loligo*'s giant axon. Dan Alkon and Erik Kandell have done much of the recent work on short-term memory and learning in marine snails. Discussions with Dr. Alkon were useful.

16. THE EYES HAVE IT

For one of the pioneer works in understanding how nerves work see H. Keffer Hartline, "Inhibition in the Eye of *Limulus,*" Journal of General Physiology 39: 651–73. Doctors Edward MacNichol and Alan Fein helped with discussions of this chapter. Also see Robert Barlow, "Limulus Brain Modulates the Structure and Function of the Lateral Eyes," *Science* 1980.

17. HEARING, WHALES, AND CIVILIZATION

For the geology of Cape Cod see Barbara Blau Chamberlain, *These Fragile Outposts* (Garden City, N. Y.: Natural History Press, 1964). For a general discussion of animal behavior see John Alcock, *Animal Behavior: An Evolutionary Approach* (Sunderland: Sinauer Associates, Inc., 1979).

For one of the classics in field studies of animal behavior see Niko Tinbergen, *the Animal in Its World* (Cambridge, Mass.: Harvard University Press, 1972). Also see Ireanus Eibl-Eibesfeldt's classic cross-cultural film on the "universal flirt display."

18. GENES AND GENERATORS, OR HOW THE LOBSTER EVOLVED HIS POSTURE

Much of this chapter is based on the work of Jella Atema at the MBL and Ed Kravitz at Harvard Medical School. Konrad Lorenz did much of the pioneering work on fixed-action patterns. For a detailed discussion of evolution and behavior see his *Evolution and Modification of Behavior*

(Chicago: Chicago University Press, 1968). Also see Melvin Konner's *The Tangled Wing*.

To counter any male bias on my part, I relied heavily on the advice and counsel of two excellent female biologists, who can be described as sociobiologists and feminists. I would like to thank Dr. Paola Borroni and Dr. Sarah Hrdy for their help and guidance. See Sarah Hrdy, *The Woman that Never Evolved* (Cambridge, Mass.: Harvard University Press, 1981).

19. WINTER ARRIVES

This chapter is based largely on personal experiences aboard Soviet fisheries research vessels, including the *R/V Blesk* and the *R/V Belegorsk*. I would like to thank the Soviet Ministry of Fish Industry and the National Marine Fisheries Service for those experiences. For descriptions of the dynamics of the Gulf Stream, and the formation and impact of warm core rings see Carl Wunsch, "Ocean Eddies," *Oceanus* 1976.

20. EGGS, ASTERS, AND URCHINS

Fertilization and the early development of the egg have fascinated biologists for centuries; see Epel, "Sperm-Egg Interactions," *Oceanus* '76 and Epel, "The Program and

Mechanisms of Fertilization in the Echinoderm Egg,"
American Zoologist 1975. Also see Stephens, "Microtu-
bules," *Oceanus* 1976. For a discussion of *kuru* see *The
Youngest Science* by Lewis Thomas.

21. COASTAL CANCER

On the East Coast much of the work on tumors in
flounder has been conducted by Robert Murchelano of
the National Marine Fisheries Service. Carol Reinisch of
Tufts University and the MBL has done much of the work
on tumors in shellfish. Their lectures and personal cor-
respondence have been very helpful. For a discussion of
DDT and ospreys see Rachel Carson's classic book, *Silent
Spring* (Boston: Houghton Mifflin, 1962) and Frank Gra-
ham, *Since Silent Spring* (Boston: Houghton Mifflin, 1970).
Also see Jella Atema, "Smelling and Tasting Underwater,"
Oceanus 1976.

22. DIVING SEALS AND THE HUMAN HEART

For an excellent description of field work see Gerald
L. Kooyman, *Weddell Seal: Consummate Diver* (Cambridge, England: Cambridge University Press, 1981). Mary
Lou Smith, chronicler of Woods Hole, history related the
surroundings of Dr. John Kanwisher's research with cor-

morants and Dr. Kanwisher more or less verified their accuracy.

EPILOGUE: THE FUTURE OF HORSESHOE CRABS

Most of this chapter was based on personal experience, daily records, and discussions with George Buckley and officials of the Food and Drug Administration. Also see John Ropes, "Data on the Occurence of Horseshoe Crabs *(Limulus polyphemus)*" in *NMFS-NEFC Survey Samples,* National Marine Fisheries Service, 1982. For a discussion of the plight of horseshoe crabs see Nishii, "About Kabutogami," (Okayama: Sanyo Printing Co., 1975).

INDEX

INDEX

Bankhead, Tallulah, 11
Barlow, Robert, 116–17
bathythermographs, 137
bay scallops, 55–56
Bdelloura candida (flatworms), 65
behavior:
    fixed-action patterns of, 120–21, 125,
      127–32
    reproductive success and, 129–31
    see also courtship and mating; defense
      behaviors; feeding; migratory behav-
      ior
Bert, Paul, 153–54
biological clock:
    in horseshoe crab vision, 117–18
    migratory behavior and, 21–22, 93
bioluminescence, 101–4
    for defense, 102
    early theories of, 102–3
    in jellyfish, 85, 101
    medical uses of, 103
    prey attracted by, 102
    for sexual attraction, 102
birds:
    chicks of, 90, 92
    diving, 154
    feeding behavior of, 89–90, 92
    fixed-action patterns in, 121, 127
    magnetic detection in, 124
    migratory behavior of, 49–50, 89, 90–93
birth:
    diving response in, 153
    of sharks, 56–57
    see also courtship and mating; eggs;
      reproduction; spawning
blood:
    cystic fibrosis and, 143
    diving response and, 152, 153, 158
    endotoxins, 72–73
    see also horseshoe crab blood; *Limulus*
      amoebocyte lysate
blue-green algae, 27–28
border discrimination, in vision, 116, 117
Boston Harbor, cleanup of, 148–49
brackish water, indicator species for, 150
bradycardia, 153
brain:
    cannibalism of, 143
    as circumesophagal ganglion, 115
    diving response and, 152, 153
    embryological development of, 106
    evolution of, 105
    of horseshoe crabs, 22, 115, 116, 117–
      18
    of lobsters, 40

neurons in, 105–6, 109–10, 128–29, 131
    receptor sites in, 99
    research on, 105–7
    vision and, 115–16, 117–18
    see also nervous system
brain diseases:
    Alzheimer's disease, 35, 110–11, 143
    microtubules and, 143
breathing, without oxygen, 152, 157, 158
breathing holes, of seals, 156–57, 158
"bug detectors," of frogs, 120

calcium:
    in cystic fibrosis, 143
    in embryological development, 141
    in generation of electrical impulses, 97
    in habituation, 109
    in human body, 103
Calidris canutus rufa (red knot), 49–50
Cambrian period, 27, 28–29, 63, 105
cancer, 146–50
    benign tumors and, 147–48
    carcinogenic pollution and, 146–50
    dietary prevention of, 147
    initiator and promotor carcinogens in,
      147–48
    molecular changes leading to, 147–48
    spread of, 148
candles (uterine pouches), 57
cannibalism, 143
Cape Cod:
    geologic formation of, 123–24
    magnetic anomaly of, 124–25
    red tide in, 96
    whale strandings on, 123–25
carcinogenic pollution, 146–50
    cleanup of, 148–49
    indicator species for, 149–50
carcinogens, initiator and promotor, 147–
    48
Carroll, Lewis, 36
Carson, Rachel, 149
catfish, 122
Cecropia moth, 113
cell division, 141, 142
cells:
    amoebocyte, 12, 74–75, 79
    carcinogenic, 147–48
    hair, 119, 120
    Mauthner, 120
    motile, as defense against infection, 57–
      58
    phagocyte, 61
    photoreceptor, 116, 118, 162
    stinging (nematocysts), 86, 99

Index

INDEX

Index

INDEX

heart diseases, bioluminescent test for, 103
heart rate, diving response and, 152, 153, 158
heredity, *see* genes, genetics
herring gulls, 127
  young, pecking response in, 121
hibernation, 136
Hippocrates, 148
Hoffstein, Sylvia, 61, 62
Homarus americanus (lobsters), 22
hormones, ecdysis regulated by, 64–65
horseshoe crab blood:
  amoebocyte cells in, 12, 74–75, 79
  bacteria in, 65–66, 74–75
  processed, *see Limulus* amoebocyte lysate
horseshoe crab eggs, 49–54, 166–67
  double shell of, 51–52
  embryological development of, 50–52
  fertilization of, 43
  genetic sex switch in, 52–54
  predators and, 49–50, 51
  spawning of, 22, 36, 42–44, 51–52, 53–54, 117
  tides and, 42–44, 51–52
horseshoe crabs, 26, 63–81, 113–18
  adolescence of, 45, 64–65
  anatomical features of, 44–45
  antibiotic system in, 65–66, 74–75
  as bait, 166
  bounties levied on, 165
  brain of, 22, 115, 116, 117–18
  of Chelicerata order, 44
  as endangered species in Japan, 166–67
  evolution of, 25, 29
  eyes of, 45, 67–68, 93, 115, 162
  feeding behavior of, 94–96, 97
  future of, 161–67
  lateral inhibition in, 116, 117, 162
  lengthy existence of, 24–25, 44
  Limulus polyphemus, 21–22
  lysate industry and, 163–64, 165, 166
  mating of, 42–45, 113, 117–18
  migratory behavior of, 21–22, 30, 36, 85, 93, 94, 118
  molting of, 45, 50–51, 64–65
  nervous system of, 21–22
  optic nerve of, 12, 115–16, 162
  population survey on, 163–64, 165
  as poultry feed or fertilizer, 165
  spawning of, 22, 36, 42–44, 51–52, 53–54, 117
  species and distribution of, 44
  trilobite larvae and, 12, 63–65

vision in, 113–18
  winter habitat of, 165
  winter hibernation of, 136
hospitals, gram-negative bacteria in, 73
Hurricane Carrie, 96

Ice Age, migratory behavior and, 89, 91–92
icefish *(Chaenocephalus aceratus),* 156
indicator species, 149–50
infections, *see* bacterial infections
inflammation, 61, 62
influenza vaccines, 78–79
initiator carcinogens, 147–48
intestines, gram-negative bacteria in, 72–73
Irish elk, 130–31

Japan, horseshoe crabs in, 166–67
Japanese squid-jigging boats, 34
jellyfish, 27, 98–101, 105
  Aequorea, 98–100, 103–4
  anatomy of, 86
  bioluminescence in, 85, 101
  comb *(Mnemiopsis),* 85–86
  equilibrium and orientation in, 100–101
  escape maneuvers of, 99–100
  feeding behavior of, 98–99, 146
  lion's mane *(Cyanea capillata),* 85, 86
  migratory behavior of, 85–87
  moon *(Aurelia aurita),* 100–101
  in neurobiological research, 103–4
  reproduction of, 86–87, 99

Kanwisher, John, 154
KGB, 96–97
Kravits, Ed, 40
Kriedl, 100
kuru, 143

larvae, 63
  eel (leptocephali), 88
  fish, in Gulf Stream, 136
  starfish, 57–58
  trilobite, 12, 51, 63–65
lateral inhibition, 116, 117, 162
lateral line system, 119–20, 121
learning:
  by habituation, 109–10
  of new motor skills, 110
leptocephali, 88
Leptonychotes weddelli (Weddell seals), 155–59
Levin, Jack, 75
light waves, 125–26

Index

INDEX

Index

INDEX

shells (egg), double, 51–52
shells (outer coverings), 28
    molting of, 36, 37–39, 45, 50, 51, 64–65,
        100
shock, endotoxins and, 73
short wave-length energy, 125
shrimp, 100
silversides *(Menida menida)*, 120
skates, magnetic detection in, 124
Skinner, B. F., 106
Smith, Langdon, 26
smooth dogfish *(Mustelus canis)*, 120, 121
    in gout research, 60, 61–62
sound waves, 125–26
    echolocation and, 122–23
south-orienting bacteria, 124
space, jellyfish in, 100–101
spawning:
    climatic conditions and, 53–54
    of eels, 88
    of flounder, 145, 146–47
    gestation vs., 56–57
    of horseshoe crabs, 22, 36, 42–44, 51–
        52, 53–54, 117
    marine vs. terrestrial, 51
    of sea urchins, 139–41, 142
    of squid, 33
    winter, 139–41, 142, 145–47
    see also courtship and mating
sperm:
    dynein and, 142
    of sea urchins, 140–41
spermatophores, 31–33, 39
spinal meningitis, 73, 80
spiny dogfish *(Squalus acanthias)*, 56–57
spring (vernal) migration, 21–23, 30–31,
    36
    see also migratory behavior
spring overturn, 23–24
spring tide, horseshoe crab eggs and, 51
Squalus acanthias (spiny dogfish), 56–57
squid, 30–35, 136
    color undulations in, 107
    commercial fishing and, 33–34, 111
    death of, 33
    defense mechanisms in, 102
    evolution of, 29
    feeding behavior of, 23, 34
    giant axons of, 34–35, 97, 108–11, 127
    habituation learning in, 109–10
    jet-propulsion of, 107–8, 127
    Loligo pealei, 22–23, 31, 97, 107–12
    mariculture facilities proposed for,
        111–12
    mating of, 31–33, 34

migratory behavior of, 22–23, 30–31
    in neurobiological research, 34–35,
        107–12
    population survey on, 165
    as prey for whales, 123
    spawning of, 33
squid-jigging boats, Japanese, 34
starfish, feeding behavior of, 55–56, 58
starfish larvae, defense against infection
    in, 57–58
stargazers, 122
startle response, 120
statocysts, 100
Stein, Gertrude, 69
Steinbeck, John, 64
sterility, in males, 142
stinging cells (nematocysts), 86, 99
stingrays:
    feeding behavior of, 87
    migratory behavior of, 87–88
    whiptail *(Dasyatis centroura)*, 87–88
stranding of whales, 123–25
    magnetic anomolies and, 124–25
stress metabolism, 153
striped bass:
    feeding behavior of, 89
    migratory behavior of, 89, 91–93
strobe lights, squid's habituation to, 109–
    10
Strongylocentrotus drobachiensis (sea
    urchins), 139–44
sudden infant death syndrome, 153, 158
sunlight, photosynthesis and, 94
super-releasing mechanisms, 129–30
swine-flu vaccine, 78–79
synapses:
    fixed-action patterns and, 128
    structural changes in, 109–10

temperature:
    embryological development and, 52
    genetic sex switch and, 52–54
    of oceans, 22–23, 28, 139
terns:
    arctic, 90–91
    feeding behavior of, 89–90, 92
    migratory behavior of, 89, 90–93
    nesting colonies of, 90
territoriality, fixed-action patterns and, 127
Tethys Sea, 44
thermocline, 137
Thoreau, Henry David, 123
Thresher, 136
tides, horseshoe crab spawning and, 42–
    44, 51–52

Index